Emerging Technologies in Hazardous Waste Management V

ACS SYMPOSIUM SERIES **607**

Emerging Technologies in Hazardous Waste Management V

D. William Tedder, EDITOR
Georgia Institute of Technology

Frederick G. Pohland, EDITOR
University of Pittsburgh

Developed from a symposium sponsored
by the Division of Industrial and Engineering Chemistry, Inc.,
of the American Chemical Society
at the Industrial and Engineering Chemistry Special Symposium,
Atlanta, Georgia
September 27–29, 1993

American Chemical Society, Washington, DC 1995

Library of Congress Cataloging-in-Publication Data

Emerging technologies in hazardous waste management V / D. William Tedder, editor; Frederic G. Pohland, editor.

   p.   cm.—(ACS symposium series, ISSN 0097–6156; 607)

   "Developed from a symposium sponsored by the Division of Industrial and Engineering Chemistry, Inc., of the American Chemical Society at the Industrial and Engineering Chemistry Special Symposium, Atlanta, Georgia, September 27–29, 1993."

   Includes bibliographical references and index.

   ISBN 0–8412–3322–5

   1. Hazardous wastes—Management—Congresses.

   I. Tedder, D. W. (Daniel William), 1946–   . II. Pohland, Frederick G., 1931–   . III. American Chemical Society. Division of Industrial and Engineering Chemistry.

TD1042.E443   1995
628.4′2—dc20
                                        95–38358
                                               CIP

This book is printed on acid-free, recycled paper.

PRINTED IN THE UNITED STATES OF AMERICA

1995 Advisory Board

ACS Symposium Series

Foreword

THE ACS SYMPOSIUM SERIES was first published in 1974 to provide a mechanism for publishing symposia quickly in book form. The purpose of this series is to publish comprehensive books developed from symposia, which are usually "snapshots in time" of the current research being done on a topic, plus some review material on the topic. For this reason, it is necessary that the papers be published as quickly as possible.

Before a symposium-based book is put under contract, the proposed table of contents is reviewed for appropriateness to the topic and for comprehensiveness of the collection. Some papers are excluded at this point, and others are added to round out the scope of the volume. In addition, a draft of each paper is peer-reviewed prior to final acceptance or rejection. This anonymous review process is supervised by the organizer(s) of the symposium, who become the editor(s) of the book. The authors then revise their papers according to the recommendations of both the reviewers and the editors, prepare camera-ready copy, and submit the final papers to the editors, who check that all necessary revisions have been made.

As a rule, only original research papers and original review papers are included in the volumes. Verbatim reproductions of previously published papers are not accepted.

Contents

Preface

HAZARDOUS WASTE MANAGEMENT continues to be of great interest to scientists and engineers. The symposium on which this volume was based featured approximately 400 presentations during a three-day meeting. The final selection of the chapters included herein is based on peer review, scientific merit, the editors' perceptions of lasting value or innovative features, and the general applicability of either the technology itself or the scientific methods and scholarly details provided by the authors.

The volume continues a theme initiated in 1990. Its predecessors, *Emerging Technologies in Hazardous Waste Management,* ACS Symposium Series No. 422 (1990), *Emerging Technologies in Hazardous Waste Management II,* ACS Symposium Series No. 468 (1991), *Emerging Technologies in Hazardous Waste Management III,* ACS Symposium Series No. 518 (1993), and *Emerging Technologies in Hazardous Waste Management IV,* ACS Symposium Series No. 554 (1994), are related contributions on waste management, but each volume is essentially different. By inspection, the reader may quickly recognize this diversity, and also conclude that no single volume can do justice to the breadth and depth of technologies being developed and applied in practice.

The contributions presented in this volume are divided into six separate but complementary sections, including: Electrokinetic Soil Cleaning; Element Recovery and Recycle; Vitrification and Thermolysis; Chemical Oxidation and Catalysis; Extraction and Precipitation; and Biological Degradation.

Acknowledgments

The symposium was supported in part by ACS Corporation Associates which is committed to excellence, solving waste problems, and reducing environmental pollution. This generosity was essential to the overall success of the symposium and is gratefully recognized.

D. WILLIAM TEDDER
School of Chemical Engineering
Georgia Institute of Technology
Atlanta, GA 30332–0100

March 14, 1995

FREDERICK G. POHLAND
Department of Civil and
    Environmental Engineering
University of Pittsburgh
Pittsburgh, PA 15261–2294

Chapter 1

Emerging Technologies in Hazardous Waste Management V

An Overview

Frederick G. Pohland[1] and D. William Tedder[2]

[1]Department of Civil and Environmental Engineering, University of Pittsburgh, Pittsburgh, PA 15261–2294
[2]School of Chemical Engineering, Georgia Institute of Technology, Atlanta, GA 30332–0100

The process of development and application of many remedial alternatives has occurred largely in response to the mandates of "Superfund" (CERCLA and SARA), as well as other federal and state programs. Hence, an array of options have become available which can include an emerging concept with bench-scale testing, a field demonstration, a selection for site-specific remediation, and full-scale use and commercialization. For example, the state-of-technology development for options applicable to soils and ground waters has been described as indicated in Figure 1(1).

Both *ex situ* and *in situ* techniques are evident, with dependence on the biological, physical and chemical sciences and engineering, either in separate or combined processes. When arrayed according to source matrix, technology applications for soils and ground waters have been or are being developed to destroy/detoxify, separate/recover, and/or immobilize contaminants(1,2).

Although not as well developed, many of these same technologies may be similarly applied to sediments, sludges and dredge spoils. Whereas, *in situ* remediation may require augmentation and/or enhancement to optimize both biological and physical-chemical techniques, *in situ* immobilization may be but one consequence of an applied technology rather than the primary intent. Moreover, techniques for access, isolation or capture of contaminants, as well as their extraction and *ex situ* aqueous-phase treatment, may provide supplemental alternatives for an integrated remedial action approach.

New technologies continue to be developed for remedial applications(3), and the use of innovative options has already surpassed established or conventional alternatives for remediation at Superfund sites. There has been a significant trend toward innovation, as new technologies are demonstrated and applied for the remediation of contaminated ground waters, soils and sediments. Moreover, depending on site-specific circumstances, more than one technology or process may be needed to achieve remediation goals. Therefore, combinations of technologies are often necessary, and those already implemented at Superfund sites have included(4,5):

0097–6156/95/0607–0001$12.00/0

- soil washing, followed by bioremediation, or incineration, or solidification/stabilization of soil fines;

- thermal desorption, followed by incineration, or solidification/stabilization, or dehalogenation to treat PCBs;

- soil vapor extraction, followed by *in situ* bioremediation, or *in situ* flushing, or solidification/stabilization, or soil washing to remove semivolatile organics;

- dechlorination, followed by soil washing for inorganics;

- solvent extraction, followed by solidification/stabilization, or incineration of extracted contaminants and solvents;

- bioremediation, followed by solidification/stabilization of inorganics; and

- *in situ* flushing followed by *in situ* bioremediation of organic residuals.

STATUS OF REMEDIAL TECHNOLOGIES
(Soils and Ground Waters)

Emerging		*Innovative*		*Established/Available*
Bench-Scale Testing Data	**Field Demonstration Use**	**Selected for Remediation**	**Limited Full-Scale Use or Limited**	**Common Full-Scale**
In situ electrokinetics	Radio-frequency heating	Solvent extraction	Thermal desorption	Incineration
X-ray treatment	*Ex situ* furnace vitrification	*In situ* soil flushing	Land treatment	Solidification/ stabilization
Electron irradiation	Pneumatic or hydraulic fracturing	Dechlorination	Soil vapor extraction	Above-ground treatment(gw)
		Bioventing	Soil washing	
Laser-induced oxidation	Treatment wall (gw)	Air sparging (gw)	*In situ* bioremediation	
			Slurry-phase bioremediation	
			In situ vitrification	

gw = ground water

Figure 1(1).

A wide range of innovative or established/conventional technologies can be identified for contaminated soils, sediments, sludges and ground waters. Many include *in situ* and *ex situ* biological, thermal, and physical/chemical processes, supplemented by techniques used primarily for containment, separation and/or enhanced recovery or off-gas treatment. Each involves a variety of challenges, including cost, performance, technical, developmental and institutional issues. Collectively, these constitute screening factors influencing the efficacy of a particular technology, ranging from overall cost to community acceptability. Such an approach has led to the development of a DOD/EPA remedial technologies matrix(5).

Remedial technology development and deployment are in considerable flux, even for those technologies ranked as having full-scale or conventional status. Moreover, there is considerable opportunity for additional discovery, using basic scientific principles and their applications in characterization of processes as well as environmental settings within which they are targeted to be applied(6,7). In this regard, some critical technology needs include:

- characterization of contaminant source matrices;

- transport and fate of contaminants in heterogeneous environments;

- solid/liquid/gas interactions;

- linking hydrometerological phenomena with geohydrologic response;

- analytical technology development;

- biological mediation;

- field verification of remedial techniques; and

- process modeling and verification.

Our present contribution in this volume to the advancement of knowledge and application in practice focuses on selected technologies currently under development, showing particular promise, or advancing fundamentals of science and technology. As such, it embellishes the continuum of topics presented in previous volumes of the series(8-11), and highlights selections that have particular contemporary relevance. Therefore, further perusal of this overview will serve to focus on technologies for hazardous waste management applied to various environmental media.

Electrokinetic Soil Cleaning

In Chapter 2, Mattson and Lindgren report on a new electrokinetic electrode apparatus designed, constructed and laboratory tested to remove water soluble chromium from unsaturated soils. Application of an electric field indicated that heavy metals such as chromium can be removed *in situ* from the unsaturated soil horizon. Krause and Tarman (Chapter 3) used a similar approach for remediating soils, and discussed primary and secondary phenomena including electroosmosis, electromigration, electrophoresis and solubility or speciation, respectively. Under conditions of constant saturation and dewatering, preliminary results suggested that increasing temperatures decreased processing time, but may reduce efficiency of extraction of dichromate from kaolinite soils.

In Chapter 4, Thornton and Shapiro developed a conceptual model for electrode designs and water treatment schemes for chromate-contaminated soils, and applied it to estimate remediation times and power consumption as functions of electrode array, chromate concentration, and applied voltage as well as capital, installation and operating costs. Lindgren, et al. (Chapter 5) also modeled elektrokinetic phenomena in unsaturated porous media, ensuring electrical neutrality, with preliminary results for a system of strong electrolytes serving as a basis for modeling more complex contaminated unsaturated soils with hydraulic flows.

Element Recovery and Recycle

Adsorption of antimony onto amorphous iron hydroxide as a treatment method for industrial wastewater was examined by Bagby and West in Chapter 6. Rapid adsorption occurred within a relatively narrow pH range, although elevated temperatures and the presence of sulfate tended to be inhibitory. Paff and Bosilovich (Chapter 7) demonstrated that secondary lead smelters could be economically used to reclaim lead from a wide range of lead-containing materials found at Superfund sites, including battery cases, lead dross and other debris containing between 3 and 70 percent lead.

In Chapter 8, Edwards, et al. present an acid treatment process for oxidizing aqueous elemental phosphorus-containing residues to produce orthophosphate slurries for subsequent reaction with ammonia to yield plant nutrient products. Removal of insoluble salts of Pb, Cd, Ba and Cr could be achieved by a solids separation step prior to conversion.

Vitrification and Thermolysis

Vitrification involves the melting and refreezing of soil to create a glass-like solid that entraps inorganic contamination, thereby isolating it from the environment. The high temperatures employed to melt the soil also destroy organic contamination. Staley (Chapter 9) discusses the performance of six vitrification technologies investigated under the U.S. EPA Superfund Innovative Technology Evaluation (SITE) Program. Reimann, et al. (Chapter 10) present developments of the iron-enriched basalt (IEB) glass-ceramic waste form at the Idaho National Engineering Laboratory (INEL) for stabilization and immobilization of large volumes of low-level nuclear wastes prior to permanent disposal. Results indicated that IEB has a high tolerance for heterogeneous waste materials, including scrap metals, while maintaining the desired chemical and physical performance characteristics. Studies with a Joule-heated melter have been extended to the application of arc furnace technology.

Experiments for thermal plasma destruction of liquid hazardous wastes such as PCBs, paint solvents, and cleaning agents have been conducted by Han, et al. (Chapter 11) to investigate optimal energy utilization while maintaining high destruction efficiency. Simultaneous production of diamond films while destroying liquid organic wastes was demonstrated at the expense of lower destruction rates and higher specific energy requirements. Possibilities of increasing destruction rates and improving specific energy requirements by reactor modifications were discussed. Czernichowski, et al. (Chapter 12) used multi-electrode plasma reactors to treat flue gases from the open burning of nitrobenzéne and trinitrotoluene. High cleaning efficiency was achieved, with substantial decreases in nitrogen oxides and carbon monoxide effluent concentrations. The use of the plasma-chemical destruction process is advocated for both soot and nitrogen oxide control.

Chemical Oxidation and Catalysis

The development of processes for destructive oxidation of organic materials has relevance to remediation of contaminated sites. Smith (Chapter 13) reports the use of air sparging with acid destructive oxidation of neoprene, cellulose, EDTA, tributylphosphate, tartaric acid and nitromethane, contrasts reaction rates with polyethylene, PVC and n-dodecane, and discusses radical formation and the mechanisms of oxidation. Patel and Vella (Chapter 14) also used air to facilitate the oxidation of sulfide from spent caustic liquors by potassium permanganate ($KMnO_4$). In the presence of oxygen, the formation of maganese dioxide provides additional oxidizing potential, thereby reducing the theoretical amount of $KMnO_4$ needed for complete sulfide oxidation.

The optimization of TiO_2-mediated solar photocatalysis of pesticide rinsates is advocated by Pugh, et al. (Chapter 15) as an attractive remediation technology for agrochemical dealers, manufacturers and farmers, and for contaminated ground waters. Studies on atrazine to determine the ideal cover material based on UV light transmittance resulted in the selection of UV transmitting acrylic with the TiO_2 bound to fiberglass mesh. The most efficient photocatalysis was achieved using five layers of mesh, a stirred reaction, water low in carbonate and other ions, a dilute waste stream, and solar irradiation rather than a mercury-vapor lamp.

Extraction and Precipitation

Lime and limestone are the most commonly used and effective chemical reagents for treating acid mine drainage. Huang and Liu (Chapter 16) report on the development of a neutralization process for treating acid mine drainage from the Berkeley Pit, one of the world's largest ore deposits containing copper and other metals. Testing included one-stage and two-stage neutralization with lime and limestone, metal ion precipitation, and copper cementation with scrap iron prior to neutralization. Multi-stage operations were considered necessary to meet U.S. EPA Water Quality Criteria for Aquatic Life (Gold Book).

Heavy metal contamination of soil is a common problem at many hazardous waste sites, with lead, chromium, cadmium, copper, zinc and mercury frequently encountered. Hong, et al. (Chapter 17) discuss the application of chelating agents for extracting heavy metals from soils, using N-(2-acetamido)iminodiacetic acid (ADA) for selective removal of zinc in the presence of competing iron and calcium ions. Chemical equilibrium modeling was used as a predictive tool for selecting suitable chelators and treatment conditions.

Cosolvents enhance the remediation of contaminated soils by mobilizing residual non-aqueous-phase liquids (NAPLs), increasing solubility, reducing sorption or retardation, and increasing mass-transfer rates. Augustijn and Rao (Chapter 18) reviewed the theoretical basis of these mechanisms, and illustrated the potential use of organic cosolvents for *in situ* remediation (solvent flushing) by several studies demonstrating enhanced contaminant removal, including for example, triethylamine (TEA), propane, supercritical carbon dioxide, alcohols and n-butylamine.

Soils and aquifers can be contaminated with viscous, nonvolatile oils such as crude oils, wood preserving chemicals and transmission fluids. For contaminants such as these, Davis (Chapter 19) advocated the use of hot water to lower viscosity and facilitate movement to recovery wells or trenches. In displacement experiments using two different soils and sands, oil recovery was enhanced as temperature increased, without injection of potentially harmful chemicals or adversely affecting indigenous microbial populations.

Biological Degradation

Although indigenous populations of microorganisms offer opportunities for *in situ* bioremediation at hazardous waste sites, genetically engineered organisms may eventually provide a more efficient approach. To create genetically manipulated microorganisms, the metabolic pathways responsible for biodegradation of toxic compounds need to be understood, and the genes encoding the appropriate enzymes must be cloned and analyzed in detail. After screening a large number of Actinomycetes for their ability to utilize a wide range of hydrocarbons as sole sources of carbon and energy, *Rhodococcus erythropolis* was selected by Lofgren, et al. (Chapter 20), and 21 mutants were isolated that were unable to use tridecane and/or tetradecane as carbon sources, but could be grouped on the basis their alkane utilization profiles. This analysis suggested that alkane metabolism in *R. erthropolis* may be different from many other bacteria, and that there may be a number of pathways and/or redundant enzymes used for growth on alkanes.

Tabak, et al. (Chapter 21) focused on the adsorption/desorption kinetics and equilibria of four polycyclic aromatic hydrocarbons in soils slurry systems. Acenaphthene, acenaphylene, naphthylene and phenanthrene were used as test substrates in respirometers constructed to quantify biodegradation kinetics, and abiotic adsorption and desorption rates, cumulative oxygen consumption, and carbon dioxide evolution were used with non-linear regression techniques to determine input parameters for a detailed simulation model.

The uncoupling of oxidative phosphorylation occurs when the energy yielding oxidation (catabolic step) is "uncoupled" from the formation of adenosene triphosphate (ATP) and the production of new cells (anabolic step). Okey and Stensel (Chapter 22) obtained multiple species uncoupling data and, using a linear free energy relationship (LAER) for the quantitative structure-activity relationship (QSAR), examined whether data from one species could be used to broaden the database for any or all species. The three different species or test systems included activated sludge, a ciliate protozoa, and plant cells grown in disperse culture. Employing the halogenated phenols, nitrophenols and anilines as test compounds in the decoupling studies with substantially different organisms, data from one species could be modeled and used with caution to estimate the biological response expected from the other species.

Summary

The management of hazardous wastes continues to command the attention of scientists and engineers in attending to existing contaminated environments, minimizing waste production at its source, and maximizing opportunities for recovery and reuse. Therefore, it is appropriate that additional attention has been directed at defining the nature of the hazard and mobilizing more efficient and innovative techniques for its reduction and elimination. Accordingly, the genesis of new and emerging technologies to cope with the complexities of impacts in often obscure and ill-defined environmental settings is expected to continue unabated for the foreseeable future.

As indicated in the introduction to this overview, both specific and combinations of technologies are often necessary to achieve a desired level of waste management and protection of health and the environment. Appropriately, focus on this need has been provided by representative examples in this volume of the series. In each case, definite progress has been achieved, but additional improvements have also been deemed necessary, whether in basic exploration of the science and/or technology or its translation into practice. Collectively, these and similar complementary efforts elsewhere continue to advance understanding and sustain progress toward resolution of both existing and future hazardous waste management challenges.

Literature Cited

1. Kovalick, W. W., "Latest News from Overseas on Innovative Technologies," Proceedings of Workshop on Innovative and Novel Technologies in Hazardous Waste Treatment and Disposal: Prospects and Problems, Second National Hazardous and Solid Waste Convention, Melbourne, Australia, May 1994.

2. Innovative Treatment Technologies: Overview and Guide to Information Sources, EPA/54019-91/002, October 1991.

3. Innovative Treatment Technologies: Annual Status Report, EPA/542-R-94-005, No. 6, September 1994.

4. Vendor Information System for Innovative Treatment Technologies (VISITT) Bulletin, EPA/542-B-93-005, July 1993.

5. Remediation Technologies Screening Matrix and Reference Guide, EPA/USAF, EPA/542-B-93-005, July 1993.

6. National Research Council, In Situ Bioremediation, National Academy Press, Washington, D.C., 1993.

7. National Research Council, Alternatives for Ground Water Cleanup, National Academy Press, Washington, D.C., 1994.

8. Tedder, D. W. and Pohland, F. G., Eds. Emerging Technologies in Hazardous Waste Management, ACS Symposium Series 422, American Chemical Society, Washington, DC, 1990.

9. Tedder, D. W. and Pohland, F. G., Eds, Emerging Technologies in Hazardous Waste Management II, ACS Symposium Series 468, American Chemical Society, Washington, DC, 1991.

10. Tedder, D. W. and Pohland, F. G., Eds. Emerging Technologies in Hazardous Waste Management III, ACS Symposium Series 518, American Chemical Society, Washington, DC, 1993.

11. Tedder, D. W. and Pohland, F. G., Eds. Emerging Technologies in Hazardous Waste Management IV, ACS Symposium Series 554, American Chemical Society, Washington, DC, 1994.

RECEIVED May 23, 1995

ELECTROKINETIC SOIL CLEANING

Chapter 2

Electrokinetic Extraction of Chromate from Unsaturated Soils

Earl D. Mattson[1] and Eric R. Lindgren[2]

[1]Sat-Unsat, Inc., 12004 Del Rey Northeast, Albuquerque, NM 87122
[2]Sandia National Laboratories, P.O. Box 5800, MS 0719, Albuquerque, NM 87185

A new electrokinetic electrode apperatus was designed, constructed and tested to removed water soluble chromium from unsaturated soils. This electrode setup was tested in two laboratory scale experiments. In the first experiment, a narrow strip of soil contaminated with chromate was placed between two electrodes. A 10 mA constant current was applied for a period of 8 days. During this time 88% of the applied chromate was removed from the anode. In a simular laboratory experiment, where all of the soil between the electrodes was contaminated with chromate, 78% of the chromium was recovered at the anode. These results indicate that heavy metals such as chromium can be removed *in-situ* from unsaturated soils by appling an electric field.

Heavy-metal contamination of soil and groundwater is a widespread problem in industrial nations. Remediation by excavation of such sites may not be cost effective or politically acceptable. Electrokinetic remediation is one possible remediation technique for *in situ* removal of such contaminants from unsaturated soils. Previous papers discussing the work performed by researchers at Sandia National Laboratories (SNL) and Sat-Unsat, Inc. (SUI) (*1-3*) focused on the transport of contaminants and dyes by electrokinetics in unsaturated soils. These experiments were conducted with graphite electrodes with no extraction system. With no extraction system, the contaminants migrating through the soil will increase in concentration at the electrode. This paper discusses a technique to remove the contaminants from unsaturated soils once they have reached an electrode.

Background

To conduct electrokinetic remediation, electrodes are implanted into the ground and a direct current is imposed between the electrodes. The application of direct current leads to two effects; ionic species in the soil-water solution will migrate to the oppositely charged electrode (electromigration), and accompanying this migration, a bulk flow of soil-water is induced toward the cathode (electroosmosis) (4). The combination of these two phenomena leads to a movement of chemical species towards one or the other electrodes.

Contaminants arriving at an electrode may potentially be removed from the soil/water by one of several methods; electroplating or adsorption onto an electrode, precipitation or co-precipitation at the electrode (with possible excavation), pumping the water near the electrode, or complexing with ion-exchange resins that are coated on the electrode. Most of the above methods are designed to operate in the saturated zone. Pumping water from a well that contains an electrode appears to be the most common electrokinetic extraction method. However, unlike groundwater in saturated soil, pore water in the vadose zone is held under tension in the soil pores (5). This tension prevents the pore water in the vadose zone from flowing into extraction wells as it does in the saturated zone. Therefore, effluent extraction techniques at an electrode proposed for saturated methods will not work in the vadose zone.

Electrode Design

Our proposed electrokinetic extraction method employs the use of a porous ceramic cup similar to that described by Shmakin (6) for mining exploration. However, unlike the open porous cup design, our technique uses a vacuum applied to the interior of the electrode as is done in a suction lysimeter. A suction lysimeter is a device that uses a porous material that exhibits a high bubbling pressure to extract pore water samples from the vadose zone. To collect a pore water sample, a vacuum is applied to the inside of the lysimeter. Depending on the magnitude of the applied vacuum and the tension of the porewater in the vadose zone about the lysimeter, the resultant pressure gradient across the porous material will either cause water to hydraulically flow into the lysimeter or to leave it. Therefore, the use of suction lysimeter technology offers a technique to control the addition and extraction of water in the vadose zone. The lysimeter design can operate at any depth and minimizes the chance of uncontrolled mobilization of contaminants that may occur with surface flooding and vadose zone well infiltration methods.

Combining electrokinetic processes with soil water sampling lysimeter technology is possible by placing an electrode inside the lysimeter. Figure 1 illustrates our design of a vadose zone lysimeter electrode. The bottom portion of the electrode is constructed of porous ceramic whereas the upper portion of the electrode is constructed of an impermeable material. The fluid between the graphite electrode and the ceramic casing is continuously purged to mitigate electrolysis reactions, sweep off gas bubbles from the electrode due to water hydrolysis, remove contaminants that enter the electrode, and to introduce

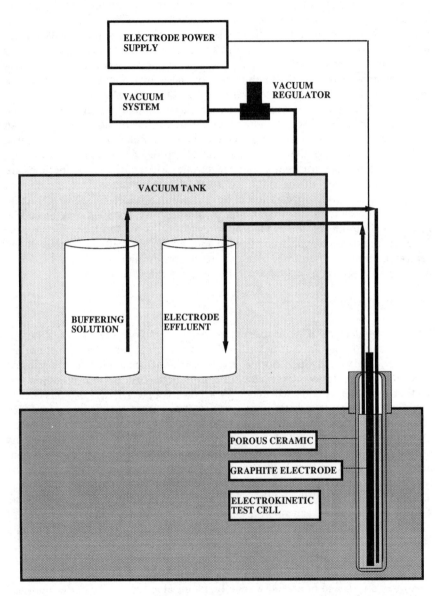

Figure 1. Lysimeter/Electrode Laboratory Design for Unsaturated Soils.

desirable ions and/or surfactants to the unsaturated soil. Solution extracted from the electrode lysimeter can be treated at the ground surface.

One problem to overcome when applying electrokinetic remediation to the vadose zone is the drying of the soil near the anode. When an electric current is applied to soil with a negative zeta potential, water will flow by electroosmosis in the soil pores, usually towards the cathode. The movement of this water will cause a depletion of soil moisture adjacent to the anode and a collection of moisture near the cathode. In SNL electrokinetic experiments conducted in a closed system (ie. no water addition or extraction) the soil near the anode dried out enough to stop the conduction of electricity. Our electrode design allows water to enter the soil at the anode to replenish the electroosmotic water, allowing continuation of the remediation process for an indefinite period. Such electrode systems have been designed for both laboratory and field applications.

A second difficulty of combining an electrode in a lysimeter is the production of hydrogen and hydroxide ions at the electrodes due to electrolysis of the water. These reactions can be neutralized by adding the appropriate acid or base, or left to build up high concentrations of hydrogen and hydroxide ions that will migrate into the soil matrix. Some researchers wish to exploit the production of hydrogen ions at the anode and propagate these ions across the soil media to strip heavy metals from the soil surfaces. Research by Acar et al. (7) have shown encouraging results, where as other research has illustrated precipitation of metal hydroxide problems occurring near the cathode and complexation changing the net charge on the ion (8,9).

Due to some of the potential problems associated with pH effects (as discussed above), and the alkaline nature of the soils in the southwestern United States, our approach is to neutralize the electrolysis reactions and replace the hydrogen and hydroxide ions with other ions. Care must be taken in choosing the replacement ions such that precipitates are not formed or the replacement ions do not react with the soil, thus changing its physical properties.

Experimental Design

The above described electrode system was tested in the laboratory to demonstrate its ability to transport and remove chromium from unsaturated soils. The experimental setup consisted of rectangular acrylic cells having internal dimensions of 3/4 x 6 x 10 inches. The cell was packed with native soil where particles greater than 16 mesh were removed. The soil was chemically equilibrated with a simulated groundwater solution (0.005 M $CaCl_2$) prior to packing.

Two experiments were conducted to test the new electrode design. The first experiment had an approximately 2 cm strip of soil contaminated with a chromate solution placed midway between the electrodes. The second experiment was a fully contaminated experiment, where a chromate solution was added to the soil everywhere between the electrodes.

In the contaminated strip experiment, the electrodes were emplaced at each end of the cell and a vacuum was applied to the electrodes to desaturate the soil. No current was applied to the cell during this portion of the

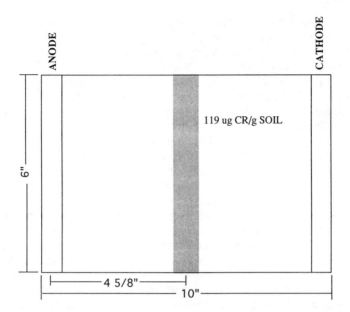

Figure 2.   Initial location of chromium in the "Contaminated Strip" experiment.

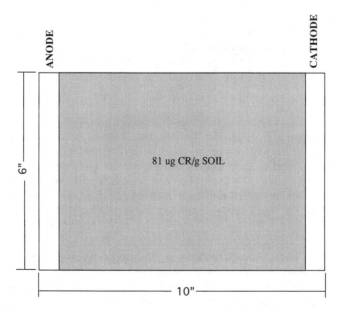

Figure 3.   Initial location of chromium in the "Totally Contaminated" experiment.

experiment. After the soil moisture came into equilibrium with the applied vacuum, a strip of soil was removed between the electrodes, microwaved dried, and contaminated with 20 ml of 1000 ppm potassium chromate solution to match the microwaved determined moisture content. This contaminated soil (119 μg Cr/g soil) was then repacked into the cell as seen if Figure 2. A constant current of 10 mA was then applied to the electrodes. The electrode reactions were neutralized with a phosphate buffer solution. Effluent from both electrodes was collected for a period of 7.8 days and analyzed for chromium by emission spectrometry. At the end of the experiment, the soil was sectioned and chemically analyzed for water soluble chromium.

The fully contaminated experiment was conducted much like the above described experiment except the soil was totally contaminated with the potassium chromate solution at a concentration of 81 μg/g soil prior to packing, Figure 3. The applied vacuum was adjusted to be in equilibrium with the soil moisture. The electrode reactions in this fully contaminated experiment were neutralized with HCl and NaOH. This experiment was operated for almost 14 days before termination. The soil was sectioned and analyzed by emission spectrometry for both water extractable chromium and acid extractable chromium.

Results

In the contaminated strip experiment, over 88 percent (17.64 mg) of the initial 20 mg chromium was recovered in the anode effluent, Figure 4. The chromium was in its anionic form as chromate ($CrO_4^=$) and was transported to the anode by electromigration. The chromate migration rate was greater than the electroosmotic flow rate that was in the opposite direction. The electromigration transport hypothesis is supported by no detection of chromium in the cathode.

As seen by the slope of the cumulative collection of the chromium verses time in Figure 4, chromium was not detected in the anode effluent until approximately 20 hours into the test. The bulk of the chromium arrived between 20 and 100 hours. After 100 hours, only a small amount of chromium was detected in the effluent at a fairly constant rate.

Destructive sampling of the soil in the test cell and subsequent water extraction of these soil samples, recovered an additional 2.5% of the applied chromium. This remaining water soluble chromium in the soil was found to be located adjacent to the anode at concentrations less than 0.22 μg Cr/gm soil, Figure 5. It is hypothesized that this remaining chromium would have entered the anode and been extracted if the system continued to be operated for a slightly longer period of time. Little chromium was detected in the initially contaminated soil strip or towards the cathode portion of the cell.

Cumulative extraction of the chromate from the anode versus time is illustrated in Figure 6 for the fully contaminated experiment. There appears to be no apparent delay in the extraction as expected. The shape of the curve indicates a fairly uniform extraction rate for the first 3 days, with decreasing extraction rate for the remainder of the experiment.

The soil in the cell was destructively sampled into 30 sections as seen in Figure 7. The water extractable chromium ($CrO_4^=$) was able to account for an

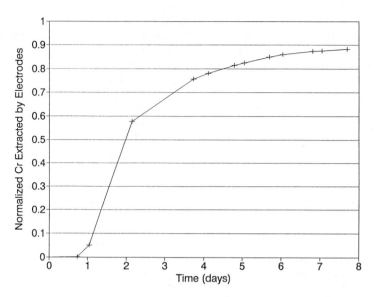

Figure 4.   Normalized cumulative extraction of chromium in the "Contaminated Strip" experiment.

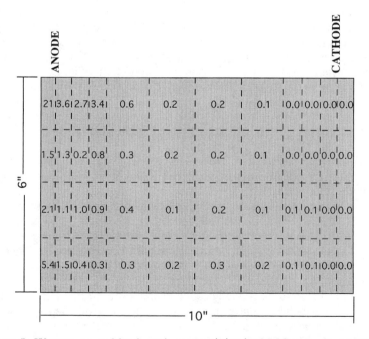

Figure 5. Water extractable chromium remaining in the "Contaminated Strip" experiment.

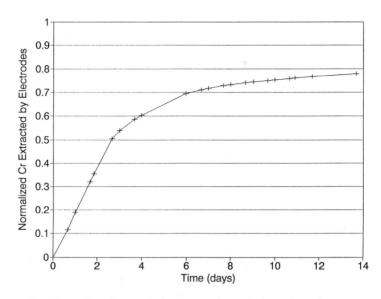

Figure 6. Normalized cumulative extraction of chromium in the "Totally Contaminated" experiment.

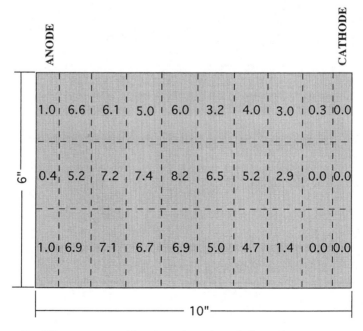

Figure 7. Water extractable chromium (μg Cr/g soil) remaining in the "Totally Contaminated" experiment.

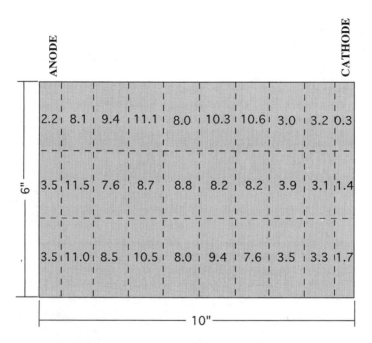

Figure 8. Acid extractable chromium (μg Cr/g soil) remaining in the "Totally Contaminated" experiment.

additional 6.1% with the highest concentrations being located near the anode. Acid extracted soil samples that were analyzed for total chromium ($CrO_4^=$ and Cr^{3+}), Figure 8, exhibit a similar relative concentration profile as the water extracted soils but shown higher concentrations. The acid extracted soils accounted for 15.9% of the applied chromium.

It is inconclusive whether the chromium remaining in the electrokinetic cell is in its trivalent form (Cr^{3+}) or if the chromium is still in its anionic form ($CrO_4^=$). Experimental conditions and electrode extraction results support the anionic form of chromium, but the greater concentrations obtained in the acid extracted soil analyses compared to the water extracted analyses suggest a change of some of the chromium to the trivalent form.

The conditions of the electrokinetic experiment were set so that is was unlikely that the chromium would have been reduced to its trivalent state. Ferguson and Nelson (*10*) stated that anionic chromium is stable in oxidizing environments with near neutral to alkaline pH values. Since the electrode reactions were neutralized and the soil has a high buffering capacity, the pH in the electrokinetic cell was such that no portion of the cell should have ever gotten below 6. In addition, the soil contains little reducing species such as Fe^{2+} or organic matter. The high pH and small reducing potential of this experiment imply that the chromium should be in its anionic state.

If the chromium had changed valences to its trivalent state, it would have migrated towards the cathode and possibly been detected in the cathode effluent. No chromium was detected in the cathode effluent, however trivalent chromium may have sorbed on the soil prior to reaching the cathode. The distribution of chromium in the electrokinetic cell after treatment (Figure 8) was such that near the cathode, less than 5% ($\{<4\mu g/g\}/\{81\mu g/g\}$) of the initial concentration remained.

The new electrodes allowed for analysis of the chromium breakthrough curves in both experiments. These analyses were performed to calculate time to extract the center of mass of the chromium. Using these times and the initial location of the center of mass, the linear migration velocity was calculated for both experiments. Surprisingly, the linear migration velocity of the chromium for both experiments was nearly identical (7.3×10^{-7} m/s for the contaminated strip experiment and 6.6×10^{-7} m/s for the fully contaminated experiment) even though the final moisture contents of the two experiments were 15.7 and 12.8% (by weight) respectively.

Summary

Uncontrolled water release to the vadose zone, long term operation, and pH conditioning of the electrode reactions were the major difficulties to overcome when designing an electrode system to extract contaminants from unsaturated soils. We developed a new unsaturated soil electrokinetic electrode system that incorporates soil physics technology and demonstrated it to be effective for removing chromium at the anode in unsaturated soils.

Laboratory bench scale experiments illustrated that a majority of the chromate can be extracted (88 and 78%) in a relatively short period of time with additional removal of chromium likely if the experiments were conducted for a longer period of time. Subsequent, destructive sampling of the soil in the cell revealed little recovery of water soluble chromium in the cell. The remaining water soluble chromium was at higher concentrations at the anode with lower to zero concentrations located near the cathode. Extraction of soil samples with nitric acid enabled a greater recovery of the remaining chromium, however these results are inconclusive concerning the valence of the remaining chromium.

The new electrode design also allows for analysis of the contaminant breakthrough curve. These analyses can be used to determine the linear electromigration velocity, and assist in determining when the experiment should be terminated.

These results are preliminary, additional tests of the electrode extraction system are being conducted to evaluate the use of different electrode neutralization solutions on the removal efficiency. Moisture content distribution, chromium valence, and electroosmosis rates are also being studied in these tests. Pilot scale electrodes have been build and are currently being tested.

Acknowledgments

This work was performed at Sandia National Laboratories which is operated for the U. S. Department of Energy under Contract No. DE-AC04-76DP00789 and funded by the Office of Technology Development, within the Department of Energy's Office of Environmental Restoration and Waste Management, under the In Situ Remediation Technology Development Integrated Program.

Literature Cited

1.   Lindgren, E.R., Kozak, M.W., and Mattson, E.D., Proceedings of the ER'91 Conference at Pasco, WA Sept 1991.
2.   Lindgren, E.R., Kozak, M.W., and Mattson, E.D., Proceedings of the Waste Management 92 Conference at Tucson, AZ, 1992.
3.   Lindgren, E.R., Mattson, E.D., and Kozak, M.W., In *Emerging Tech. in Hazardous Waste Management IV ACS Symposium Series*; Tedder, D.W., and Pohland, F.G., Ed.; Series 554, ACS, Washington, DC, 1994.
4.   Hunter, R.L., *Zeta Potential In Colloid Science*, Academic Press, New York, NY, 1981.
5.   Hillel, D., *Fundamentals of Soil Physics*, Academic Press, NY, NY, 1980.
6.   Shmakin, B.M., *J of Geochem. Explor.*, 1985, 23.
7.   Acar, Y.B., Gale, R.J., Putnam, G., and Hamed, J., Proceedings of the 2nd Int. Sym. on Envir. Geotech., Envo Publishing, Bethlehem, Pa, 1989.
8.   Hamed, J., Acar, Y.B., and Gale, R.J., ASCE, *J of Geotech. Eng.*, 1991, 117(2):241-271.
9.   Runnells, D.D., and Wahli, C., *Ground Water Monitoring Review*, Winter, 1993.
10.   Ferguson, J.F., and Nelson, P., Unpublished proceedings of the Electrokinetic Treatment and its Application in Environmental - Geotechnical Engineering for Hazardous Waste Site Remediation, Seattle, WA, 1986.

RECEIVED March 14, 1995

Chapter 3

Preliminary Results from the Investigation of Thermal Effects in Electrokinetic Soil Remediation

T. R. Krause and B. Tarman

Chemical Technology Division, Argonne National Laboratory, 9700 South Cass Avenue, Argonne, IL 60439–4837

Electrokinetics is an emerging soil remediation technology. Contaminants are extracted from the soil as a result of a complex set of phenomena that occur when an electric gradient is imposed across a soil-water system. The primary phenomena include electroosmosis, electromigration, and electrophoresis. Secondary phenomena, such as changes in solubility or speciation of various chemical components, may occur as a result of electrically induced changes in the chemical environment of the system. Numerous factors, such as temperature, may affect each of these phenomena and, consequently, the overall process efficiency. We have begun an investigation of thermal effects in the extraction of potassium dichromate from kaolinite soils under conditions of constant saturation and dewatering. Preliminary results suggest that increasing the soil temperature from 21 to 55 °C may decrease the processing time under saturated conditions. However, increasing the soil temperature under dewatering conditions causes soil cracking, which reduces the overall process efficiency.

Soils contaminated with heavy metals and organic compounds is a major environmental concern at the Department of Energy - Defense Production (DOE-DP) facilities. For example, soils contaminated with chromium can be found at Sandia National Laboratory, with mercury at the Oak Ridge Reservation, and with process waste, including TRUs, strontium, and cesium, from leaking storage tanks at the Hanford Site. The Office of Technology Development (OTD) within DOE is responsible for developing remediation and waste management technologies to resolve or minimize the environmental impact of these contaminated soils. One remediation technology which has received considerable attention is electrokinetic remediation.

"Electrokinetic remediation" occurs when several transport phenomena collectively promote contaminant migration in a soil-groundwater system as a result of an applied electric gradient between electrodes embedded in the soil subsurface. These phenomena include electroosmosis, electromigration, and electrophoresis. Electroosmosis is the flow of groundwater. Electromigration and electrophoresis

0097–6156/95/0607–0021$12.00/0

describe the movement of charged species, ions or colloidal particles, respectively, relative to the fluid phase. Decontamination of a site results from the controlled transport of a contaminant either by convective forces due to electroosmosis or by transport mechanisms due to electromigration or electrophoresis from the contaminated region to a collection region. For metal species, generally, electromigration is the dominant force (1).

At the molecular level, a complex and interrelated set of chemical and physical phenomena are observed during electrokinetic processing. The phenomena effecting the contaminant(s) to be removed include convective and diffusional transport processes, sorption from the soil, and chemical reactions resulting in changes in the compound or, in the cases of metal ions, changes in oxidation state and/or ligand coordination sphere. In addition, changes occur in the chemical and physical properties of the soil-water interface and the effect of the biological properties of the soil play a significant role in the efficiency of the electrokinetic process. The overall efficiency of the transport processes will depend on many factors, including the ion concentration, the hydration of ions and charged particles, the mobility of the ions and charged particles, the viscosity of the groundwater, the dielectric constant of the medium, and the temperature (2).

The electrokinetic phenomena, in particular, electroosmosis was first observed by Reuss in 1808 (3). The initial application of this phenomena was electroosmotic dewatering and hence, stabilizing, soils and sludges to permit construction of building foundations, earth-filled dams, airport runways, and railroad right-of-ways (4). In 1980, Segall and co-workers (5) reported the detection of high concentration of metals and organic compounds in water electroosmotically drained from a sludge. Since then a considerable amount of experimental and theoretical work aimed at developing this technology has been performed in academic research laboratories (1,6-19) government laboratories (20-21), and industrial laboratories (22-23). In the United States, current field applications of electrokinetic remediation has been limited to an EPA-sponsored demonstration (24). In Europe, a Dutch firm, Geokinetics has performed several field-scale attempts to remediate heavy metals from saturated soils (22).

Heating of soil during electrokinetic remediation can result from two mechanisms: intentional heating of the soil using technologies such as imbedded heating elements or radio-frequency heating or unavoidable heating as a result of resistive heating due to high current/voltages during processing. Currently, there is little information available in the literature discussing the effect of temperature in electrokinetics processing. Winterkorn (25) investigated the effect of resistive heating of the soil during electrokinetic processing. He suggested that a thermal gradient term must be included with the hydraulic and electroosmotic gradients to predict the electroosmosis flow to account for resistive heating of the soil. Tarman (26) investigated thermal heating to enhance the recovery of chlorinated hydrocarbons from various fine-grained soils. Due to the complex interrelationship between transport properties, sorption processes, and chemical reactions, which may occur during electrokinetic processing, it is very difficult to predict a priori how increases in the process temperature will effect the overall electrokinetic process efficiency (10). As a consequence, most research is conducted under low current/voltage conditions to minimize resistive heating of the soil.

Thermal effects on electrokinetic remediation of chromate-laden kaolinite under conditions of both constant saturation and dewatering were investigated. Although considerable academic research has been conducted on fundamental processes that occur during electrokinetic remediation under conditions of constant soil saturation, there have been limited research under dewatering conditions. However, since a number of potential sites that are a candidate for application of this technology are located in arid region, a significant amount of research conducted within DOE has focussed on electrokinetic processing under unsaturated conditions *(20-21)*.

Experimental

Chromate-laden soil samples were prepared by mixing an equivalent weight of a 1000-ppm chromium solution (0.0096 M $K_2Cr_2O_7$) with kaolinite (Thiele Kaolin Co.). The slurry was mechanically mixed to a uniform consistency. The slurry was loaded into the sample cell, an acrylic tube 3.5 cm ID and 8.5 cm long. The slurry was then consolidated to drain excess water by applying pressure in incremental steps of 3-5 psi up to a total pressure of 30-40 psi over a 24-h period *(18)*. After consolidation, two porous stones were glued with a silicon-based epoxy to each end of the sample cell to hold the soil in place. The sample cell was placed in the electrokinetics test apparatus provided by Lehigh University shown in Figure 1 for processing. Using this procedure, samples could be consistently prepared samples with an initial degree of saturation ranging from 90-100%, indicating nearly all of the pore volume contains water, and a chromium concentration in the range of 0.04-0.06 wt% (based on the dry weight of soil). A number of baseline samples containing no chromium were prepared by a similar procedure.

Electrokinetics processing conditions was investigated in "open" and "closed" systems *(4)*. In the open-system, water reservoirs were provided at both the anode and cathode electrodes to replace water loss due to electroosmosis and maintain the saturation level of the soil at 85-100% during processing. In the closed-system, no water reservoirs were provided and the soil was allowed to dewater, as a result of electroosmosis (i.e., the saturation level continuously decreased).

Electrokinetics processing was conducted at a constant d.c. voltage of 30 v. (Kenmore Model PD110-3D or PD110-5D Power Supplies). Small graphite electrodes, approximately 0.25 in. (0.64 cm) long and 0.06 in. (0.15 cm) in diameter, were inserted into the soil at various locations to measure intermediate voltage drops along the length of the cell. Current and voltage measurements were made periodically using a Fluke Model 8060A True RMS Multimeter. In the open system, the electrodes were constructed from four graphite rods, approximately 2.25 in. (5.72 cm) long and 0.25 in. (0.64 cm) in diameter, which were glued together. The electrodes were submersed in the water reservoirs. In the closed system, the porous stones, which served to hold the soil samples in the test cell, served as the electrodes. Electrical connections were made by gluing a lead wire from the power supply to the surface of the porous stone using a conductive carbon cement (Neubauer Chemikalien).

Experiments performed at elevated temperatures, 42 and 55 °C, were conducted by wrapping the cells with heating tapes. Temperature was measured at three locations

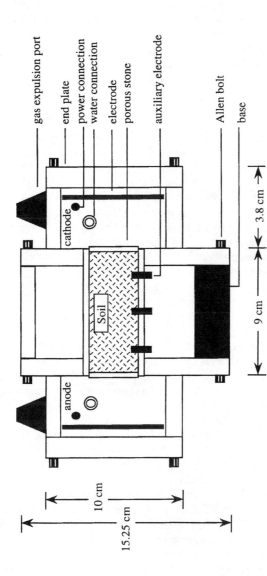

Figure 1. Electrokinetic cell provided by Lehigh University. Reproduced with permission from ref. 30.

using chromel-alumel thermocouples inserted into the soil. The thermocouple located at the midpoint between the anode and cathode served as the reference thermocouple for a Omega Temperature Controller (Model CN320) that controlled the power supplied to the heating tapes. The samples were heated to the desired temperature and allowed to equilibrate for approximately 30-60 min. before the voltage was applied. During a typical experiment, the temperature varied by about 1-2 °C across the cell. In the open system, the water in the reservoirs was maintained at ambient temperature, 21 °C.

In open-system processing, small alliquots were periodically removed from the cathode and anode reservoirs. In closed-system processing, small soil samples were recovered after processing. Quantitative analysis of these alliquots and the soil samples for chromium was performed by the Analytic Chemistry Laboratory. Ion-coupled plasma spectroscopy was used on liquid samples. A modification of standard EPA methods (Series 3000 Methods) were used on soil samples for chromium determination. No determination of chromium speciation was attempted during these initial tests.

Results and Discussion

Open system. Open system experiments were conducted at three temperatures, 21, 42, and 55 °C, for a period of 7 hours. Three samples were processed at each temperature to indicate the reproducibility of the results.

Since chromium is in an anionic form, either $Cr_2O_7^{2-}$ or CrO_4^{2-}, under conditions generated by the electrokinetic processing, it will migrate due to the potential difference towards the anode. Electromigration is the predominant transport process. Increasing the processing temperature above ambient conditions (21 °C) resulted in an increase in the total amount of chromium accumulation in the anode electrode reservoir during the 7 hour tests, as illustrated in Figure 2. During the first 1-2 hours, rate of accumulation of chromium did not appear to vary with temperature. However, after this initial 2 hour period, the rate of accumulation of chromium did increase with increasing process temperatures, although there was no difference observed in the rate of accumulation at the above ambient temperatures (42 and 55 °C) investigated. A small amount of chromium did accumulate in the cathode chamber either due to convective transportation by the groundwater or to diffusion caused by concentration gradients. As shown in Figure 3, the amount of chromium accumulated in the cathode increases with increasing process temperature. Based on the concentration of chromium in the anode chamber, approximately 20% of the chromium initially present in the soil sample was recovered at 21 °C over a 7 hour period. Approximately 35% of the chromium initially present in the soil was recovered at 42 and 55 °C over the same time period. No attempt was made to recover any more chromium.

The volumetric flow rate of water into the anode and out of the soil at the cathode were measured. In general, the total volume of water transported across the soil sample increased with increasing temperature, as shown in Figure 4. At the highest temperature investigated, 55 °C, the amount of water entering and leaving the soil sample are approximately equal. However, this amounted to only about 10-15%

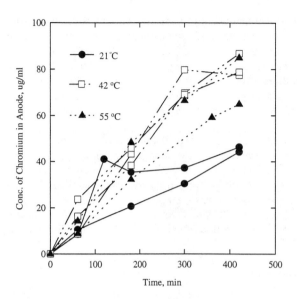

Figure 2. Concentration of chromium that accumulated in the anode chamber at 21, 42, and 55 °C using the open system. Reproduced with permission from ref. 30.

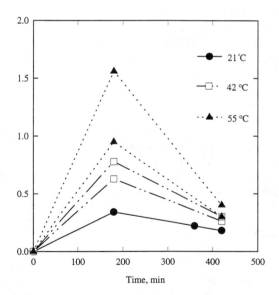

Figure 3. Concentration of chromium that accumulated in the cathode chamber at 21, 42, and 55 °C using the open system.

of the total volume of water initially present in the soil. Thus, although there is a net imbalance of flow across the samples at lower temperatures, the small amount of flow, (less than 5% of the initial volume of water present in the soil) and the short processing times did not result in a significant change in the degree of saturation of the soil. Similar observations were observed using blank samples, which contained no chromium, under similar processing conditions.

Another system parameter that was measured was the current. The current did not appear to be a function of the temperature as illustrated in Figure 5. As discussed earlier, these experiments were performed at a constant d.c. voltage of 30 v. maintained between the two electrodes. Most of the voltage drop, > 80-90%, was across the soil samples.

In general, chromium electromigration was the dominant transport mechanism. Transport processes, such as convective transport and diffusional processes did not substantially retard the electromigration of chromium. This observation is consistent with results obtained for chromium removal from kaolinite *(18)* and sand *(21)* at ambient temperatures. Increasing the process temperature above ambient temperature did increase the rate of chromium removal from the soil by nearly a factor of two. Not enough experiments were conducted to determine the effect of temperature at lower initial chromium concentrations and/or saturation conditions below 100% in the open system.

Closed system. A series of experiments were conducted under dewatering conditions at the same three temperatures investigated in the open system, 21, 42, and 55 °C. Three separate samples were prepared for each temperature, and processed for 3, 7, or 24 hours. The reproducibility of the results was determined by comparing measured parameters, such as voltage drops and current, of different samples for similar processing times (e.g. the first three hours of processing time of the 7- and 24-h experiments compared to the entire 3-h experiment).

The chromate-laden kaolinite samples were prepared as already described for uniformity. However, water reservoirs were not utilized. The porous stones, which are in direct contact with the soil, served as the cathode and anode. During this operation, dewatering of the kaolinite samples occurred due to electroosmosis, thermally-induced stress dewatering, and evaporation. Chromate and water were allowed to flow from the soil through the porous stones and into a collection basin. No attempt was made to quantify the chromate or the water extracted from the samples. Only the initial and final chromium concentrations of the soil samples were measured. Based on the analytical results, the amount of chromium removed ranged from approximately 50% of the initial chromium concentration at 55 °C to >80% at 42 °C for samples processed for 3 h.

As shown in Figure 6, the chromium concentration is greatest in the vicinity of the anode for all samples at all temperatures investigated. Chromium concentrations in the vicinity of the cathode were significantly below the initial chromium concentrations. As in the open system, this observation is consistent with the electromigration of the negatively-charged chromate moiety towards the anode, opposite the electroosmotic flow of water towards the cathode. The amount of chromium present near the center of the test cell was found to increase with increasing temperature.

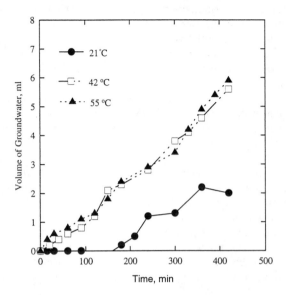

Figure 4. Volumetric flow of groundwater into the cathode chamber at 21, 42, and 55 °C under saturated conditions using the open system.

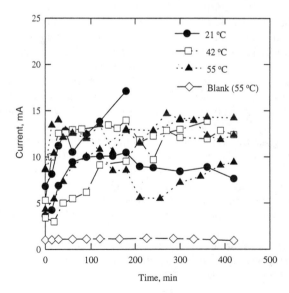

Figure 5. Observed current for a constant d.c. voltage of 30 v. at 21, 42, and 55 °C under saturated conditions using the open system.

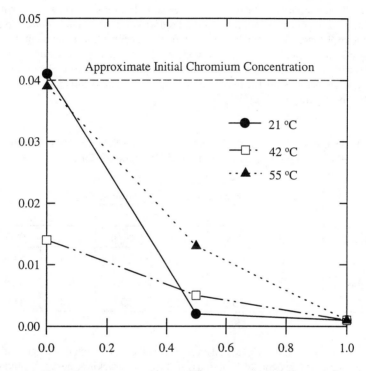

Figure 6. Chromium concentration profile versus relative distance from anode after 3 h at at 21, 42, and 55 °C using the closed system. Reproduced with permission from ref. 30.

Chromate-laden water was observed to drain from both the anode and cathode regions of samples during the time required to heat the samples to 42 and 55 °C. Qualitatively, the amount of water that drained from the samples increased with increasing processing temperature. During this period, no electric gradient was applied to the soil system, suggesting that thermally-induced dewatering occurred.

As anticipated, the percent loss of water increases with both increasing temperature and duration of the experiment, as shown in Table I. No attempts were made to distinguish between water loss due to electroosmosis and thermal effects, such as evaporation or thermally-induced stresses (27). Our observation that the ability to move chromate anions decreases as the soil dewaters is consistent with an investigation of the electrokinetic transport of chloride anions in sand conducted at the University of Manchester which observed that it became more difficult to move chloride anions in sand as the water content decreases (2). Previous results for chromate and dye removal from sand suggests that the maxim electromigration velocity can be maintained for moisture contents well below total saturation before a decrease in mobility is observed (28).

Table I. Degree of Saturation (Percent) of Kaolinite Samples after Various
          Processing Times as a Function of Temperature for a Closed System

| Temp(°C) | Percent Saturation of Kaolinite Samples | | | | | |
| | 3 h | | 7 h | | 24 h | |
	Initial	Final	Initial	Final	Initial	Final
21	95.4	94.7	96.1	92.5	95.9	75.7
42	94.5	76.4	100	73.5	102	60.5
55	95.1	73.9	94.6	60.4	97.8	31.9

Cracks, due to soil shrinkage, developed in all cases for experiments conducted at elevated temperatures for 24-h periods. In fact, the cracking, hence shrinking, of the soil samples was so severe for the experiment conducted at 55 °C, that the current decreased by three orders of magnitude. Cracking causes a discontinuity of the pathways for contaminant transport. As pointed out by Casagrande, shrinkage will continue to the plastic limit, and depending on the applied voltage, to the shrinkage limit. Once this limit is reached, electroosmosis (and hence, electrokinetics remediation) will stop (29). As a consequence, electrokinetics remediation under dewatering conditions for kaolinite soils will be difficult, especially as the degree of saturation decreases.

Conclusions

In summary, due to the limited number of experiments performed, no concrete conclusions can be derived. However, based on a number of observations, some suggestions can be made to assist or provide guidance for future work in this area and in electrokinetics processing in general. The primary observations are as follows: (1) Increased processing temperature could result in shorter time for site remediation under saturated conditions. (2) Significant loss of groundwater during closed system operations could hinder the transport of metal species through the soil resulting in a potential redistribution of the contaminants but not necessarily decontamination of the site. (3) Thermally-induced dewatering could cause migration of contaminants along thermal gradients, which may or may not coincide with electric gradients (initial expansion of contamination could actually be observed as the processing temperature increases).

Acknowledgements

I would like to acknowledge the following individuals: Dr. Sibel Pamucku of Lehigh University for providing the electrokinetics test cells; and Dean Bass, Doris Huff, Edmund Huff, and Florence Smith of the Analytical Chemistry Laboratory at Argonne National Laboratory. This work was supported by the U. S. Department of Energy, Office of Technology Development, under Contract W-31-109-Eng-38.

Literature Cited

1. Pamukcu, S.; Whittle, J. K. In *Handbook of Process Engineering for Pollution Control and Waste Minimization*, Wise, D. L.; Trantolo, D. J., Eds.; M. Dekker: New York, NY, 1994.
2. Cabrera-Guzman, D.; Swartzbaugh, J. T.; Weisman, A. W. *J. Air Waste Manage. Assoc.*, **1990**, *40*, pp. 1670-1676.
3. Reuss, F. F. *Memoires de la Societe Imperiale des Naturalistes de Moscou*, **1809**, 2, pp. 327-347.
4. Casagrande, L. *Boston Soc. Civ. Eng.*, **1952**, *69*, pp. 255 -271.
5. Segal, B. A.; O'Bannon, C. E.; Matthias, J. A. *J. Geotech. Eng. Div., Am. Soc. Civ. Eng.*, **1980**, *106(GT10)*, pp. 1143-1147.
6. Acar, Y. B.; Gale, R. J.; Putman, G.; Hamed, J.; Wong, R. *J. Envir. Sci. and Health, Part (a): Envir. Sci. and Engrg.*, **1990**, *25(6)*, pp. 687-714.
7. Acar, Y. B.; Gale, R. J.; Putnam, G. *Transportation Research Record*, Transportation Research Board, National Research Council, Washington, DC, 1990; No. 1288, pp. 23-34.
8. Hamed, J.; Acar, Y. B.; Gale, R. J. *J. Geotech. Eng. Div., Am. Soc. Civ. Eng.*, **1991**, *117(2)*, pp. 241-271.
9. Acar, Y. B.; Li, H.; Gale, R. J. *J. Geotech. Eng. Div., Am. Soc. Civ. Eng.*, **1992**, *118(11)*, pp. 1837-1852.
10. Acar, Y. B.; Alshawabkeh, N. A.; Gale, R. J. *Waste Management*, **1993**, *13(2)*, pp. 141-150.
11. Renauld, P. C.; Probstein, R. F. *Physicochem. Hydrodyn.*, **1987**, *9(1/2)*, pp. 345-360.
12. Shapiro, A. P.; Renauld, P. C.; Probstein, R. F. *Physicochem. Hydrodyn.*, **1989**, *11(5/6)*, 1989, pp. 785-802.

13. Shapiro, A. P.; Probstein, R. F. *Environ. Sci. Technol.*, **1993**, *27(2)*, pp. 283-291.
14. Probstein, R. F.; Hicks, R. E. *Science,* **1993**, *260,* pp. 498-503.
15. Pamukcu, S.; Khan, L. I.; Fang, H. F. *Transportation Research Record,* Transportation Research Board, National Research Council, Washington, DC, 1990; No. 1288, pp. 41-46.
16. Pamukcu, S.; Wittle, J. K. *Environmental Progress*, **1992**, *11(3)*, pp. 241-250.
17. Pamukcu, S.; Wittle, J. K. Proceedings of the 14th Annual DOE Low-Level Radioactive Waste Management Conference, Phoenix, AZ, November 18-20, 1992.
18. Wittle, K. K.; Pamukcu, S. *Electrokinetic Treatment of Contaminated Soils, Sludges, and Lagoons*; Argonne National Laboratory: Argonne, IL, 1993; DOE/CH-9206.
19. Gray, D. H.; Mitchell, J. K. *J. of the Soil Mech. and Found. Div.*, *Am. Soc. Civ. Eng.*, *1967*, *92(SM 6)*, pp. 209-236.
20. Lindgren, E. R.; Kozak, M. W.; Mattson, E. D. In *Proceedings of the 1991 DOE Environmental Restoration Conference*; Wood, D. E., Ed.; U. S. Department of Energy: Washington, DC, 1991, pp. 151-157.
21. Lindgren, E. R.; Kozak, M. W.; Mattson, E. D. In *Waste Management '92*; Post, R. G., Ed.; University of Arizona: Tucson, AZ, 1992, pp. 1309-1314.
22. Lageman, R. Presented at the NATO/CCMS Pilot Study: Demonstration of Remedial Action Technologies for Contaminated Land and Groundwater, Copenhagen, Denmark, May 8-9, 1989.
23. Lageman, R.; Pool, W.; Seffinga, G. *Chem. Ind. London*, **1989**, *18*, pp. 575-590.
24. Bannerjee, S.; Horng, J.; Ferguson, J. F.; Nelson, P. O. *Field-Scale Feasibility Study of Electro-Kinetic Remediation;* U.S. Environmental Protection Agency, U. S. Government Printing Office: Washington, DC, 1988, CR 811762-01.
25. Winterkorn, H. F. *Nat. Sci.-Natl. Research Council, Publication No. 359*, Highway Research Board, National Research Council, Washington, DC, 1955; No. 108, pp. 1-24.
26. Tarman, B. M. S. Thesis, Lehigh University, Bethlehem, PA, 1992.
27. Campanella, R. G.; Mitchell, J. K. *J. Soil Mech. Found. Div. E, Am. Soc. Civ. Eng.*, *1968*, *93(SM3)*, pp. 709-734.
28. Lindgren, E. R.; Kozak, M. W.; Mattson, E. D. In *Proceedings of the Information Exchange Meeting on Waste Retrieval, Treatment, and Processing*; U. S. Department of Energy, U. S. Government Printing Office: Washington, D. C., 1993; Conf-930149, pp. 447-453.
29. Casagrande, L. *J. of the Soil Mech. and Found. Div.*, *Am. Soc. Civ. Eng.*, *1967*, *93(SM 6)*, pp. 1385-1387.
30. Battles, J. E.; Myles, K. M.; Laidler, J. J.; Green, D. W. *Chemical Technology Division Annual Technical Report*, Argonne National Laboratory: Argonne, IL, 1993, ANL-93/17.

RECEIVED June 9, 1995

Chapter 4

Modeling and Economic Analysis of In Situ Remediation of Cr(VI)-Contaminated Soil by Electromigration

Roy F. Thornton and Andrew P. Shapiro

General Electric Corporate Research and Development, P.O. Box 8, Schenectady, NY 12301

In order to estimate the economics of remediation of metal-contaminated soil by electromigration, a conceptual model of a site and preliminary engineering approaches were proposed. Electrode designs and a water-treatment scheme were proposed for a 30 m x 30 m x 2 m bed of chromate-contaminated soil. A mathematical model was developed and used to estimate remediation times and power consumption as functions of several variables including number of electrodes, chromate concentration, and applied voltage. The model assumed a constant typical soil resistivity and no adsorption of the chromate ions. The system design and outputs from the model were then used to estimate the capital, installation, and operating costs of purging the soil of the hexavalent chromium. The total remediation cost for a 30 m by 30 m by 2 m volume of soil contaminated with 100 μg/mL Cr(VI) in the groundwater was in the range of $130/ton.

In recent years much research has been conducted on the application of electrokinetic phenomena and ionic migration to soil remediation[1-4]. While several authors have reported modest energy costs for their studied processes, most have based their costs on one-dimensional experiments or analysis. In actual field applications the most common electrode shape will be a long cylinder placed in vertical or horizontal wells. When using cylindrical electrodes there will be significant nonuniformities in the applied electric field which are absent in one-dimensional experiments. Probstein and Hicks[3] presented experiments and modeling that illustrate these three-dimensional effects in electroosmotic purging. Because these nonuniformities can be very large, it is important to understand how they will affect the economics of the process. In this paper we investigate the two-dimensional effects of using cylindrical electrodes in a variety of geometries on remediation times, energy requirements, and overall costs for remediation with electric fields. As a model problem we consider removing soluble chromate by ionic migration from an 1800 m³ contaminated site.

0097–6156/95/0607–0033$12.00/0

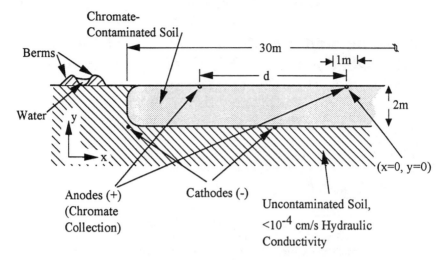

Figure 1.   Vertical cross-section of model site.

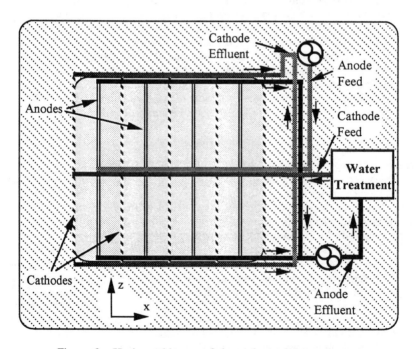

Figure 2.   Horizontal layout of electrodes and water system.

The objective of this study was to estimate the cost of remediating metal-contaminated soil using electrokinetic methods. In lieu of the time and expense of selecting and characterizing an actual site, an idealized model site was developed. A system of electrodes, methods of installation, and water treatment methods were proposed; and the remediation performance of various electrode configurations was mathematically modeled. Data from the mathematical modeling, the proposed site configuration, and engineering considerations were presented to the Brattle Group, Cambridge, MA to use in an economic analysis. Their analysis included best estimates of all costs from mobilization onto the site to demobilization after remediation. Costs common to other remediation methods such as site characterization were not included.

Methods

Remediation site concept. As a basis for estimating the cost of electrokinetically remediating soil contaminated with soluble ionic compounds, the following model site was proposed:

- A zone of chromate-contaminated soil at the surface, 30 m by 30 m by 2 m deep
- Saturated soil, or soil capable of being saturated by irrigation
- Zero groundwater velocity by reason of zero gradient or low permeability

The proposed electrode configuration was to be two horizontal layers of pipes or rods, one at the surface of the ground and one just beneath the contaminated zone. The arrays shown in Figures 1 and 2 are staggered. Also considered was a set of two arrays in which the anodes were aligned directly over the cathodes. In order to avoid driving chromate deeper into the soil in some areas, the anodes must be at the surface and the cathodes just beneath the contaminated zone. The cathodes are proposed to be iron pipes inside perforated plastic pipe laid horizontally in gravel at the bottom of two-meter deep back-filled trenches. Geotextile would be used to prevent soil infiltration into the plastic pipe. The concept is illustrated in Figure 3a. The anodes are proposed to be disposable carbon rods inside perforated plastic pipe in shallow back-filled trenches as shown in Figure 3b.

The action of the direct current electric field would be to drive chromate and other soil anions toward the anodes where they would be removed by the water flowing in the pipes. Water electrolysis at the anodes would produce oxygen gas and hydrogen ions, and the pH would drop continuously if not compensated for with hydroxyl ions. Cations would migrate toward the cathodes where they would be removed by the flowing catholyte. Water electrolysis at the cathodes would produce hydrogen and hydroxyl ions, leading to a rising pH. Water treatment would include passing the anolyte through ion exchange resins to remove the anions purged from the soil. Acid would be added as necessary to compensate for the hydroxyl ions released by the anion exchange resin. The pH of the system should be maintained above 8 to minimize chromate adsorption to the soil and below 11.5 to minimize hydroxyl ion transport into the soil from the catholyte. Most or all of the neutralizations required could be accomplished by pumping the decontaminated acidic anolyte to the cathode pipes and the catholyte to the anode pipes. Additional acid or base might be needed to control pH, and salt could be added, or water removed and replaced, to control salinity.

Mathematical Model. Calculation of remediation times and energy requirements for electrokinetic processes can be a formidable task. The main complexity stems from the transient behavior of soil conductivity which is affected by electrode reactions and chemical equilibria[2,5]. In this paper we are interested in an estimate for the purposes of economic comparison and therefore we assume the soil conductivity has a constant characteristic value. Given this assumption the electric field in a homogeneous medium is given by the Laplace equation

$$\nabla^2 \phi = 0 \qquad (1)$$

where ϕ is the electric potential. This equation has been solved analytically for many geometries including two-dimensional representations of the electrode configurations examined here[6].

Solution Technique. The remediation times and energy costs for the electrode and placement geometries considered in this paper are determined by solving for f by superposition of solutions for the Laplace equation for individual line sources. Line sources are used to represent long cylindrical electrodes whose radii are much small than the inter-electrode spacing. For an infinite array of electrodes running parallel to the z-axis, the plane of which is parallel to the xz-plane, the potential is given by

$$\phi = \frac{I}{4\pi\sigma L} \ln\left(\frac{\cosh(2\pi(y-y0)/d) - \cos(2\pi(x-x0)/d)}{2} \right) \qquad (2)$$

where each electrode carries a current I per length L (in the z-direction), d is the horizontal interelectrode spacing as shown in Figure 1, and the (x0, y0) is the position of one electrode in the array.

The velocity field for an ion in solution is given by

$$\mathbf{u} = \frac{v}{\tau^2} zF\nabla\phi = m\nabla\phi \qquad (3)$$

where τ is the tortuosity, v is the mobility in [m mol s^{-1} N^{-1}], z the charge number of the ion, and F is the Faraday constant (96,500 C mol^{-1}). The apparent mobility of the ion in the porous media, m, has units of [m^2 V^{-1} s^{-1}]. In the present problem ionic migration is assumed to be much faster than electroosmosis and therefore electroosmosis is not considered. However Equation 3 can be adapted for electroosmotic velocity by substituting the proper electrokinetic coefficients for the ion mobility.

To calculate remediation times it is necessary to determine the trajectory, or streamline, of ions in the region to be remediated. The time required for remediation is determined by integration along these streamlines. In this paper the remediation time for a given geometry is defined as the time by which at least 99% of the ions initially in the contaminated zone reach the sink electrode. In the cases analyzed here, it is mathematically simpler to calculate the velocities from the stream function j which is a constant along a streamline or current path. The stream function is related to the potential function by the Cauchy-Riemann condition:

$$\frac{\partial\varphi}{\partial y} = \frac{\partial\phi}{\partial x}, \quad \frac{\partial\varphi}{\partial x} = -\frac{\partial\phi}{\partial y}$$

The corresponding stream function for this array is

$$\varphi = \frac{I}{2\pi\sigma L} \tan^{-1}\left(\frac{\tanh(\pi(y-y0)/d)}{\tan(\pi(x-x0)/d)} \right) \qquad (4)$$

Electrode Geometries. The two electrode geometries considered were 1) two vertically aligned horizontal arrays with the top array on the soil surface and the bottom array just beneath the contaminated region, and 2) two staggered horizontal arrays such that the electrodes of the bottom array are positioned halfway between the electrodes of the top array. For the aligned arrays the potential is composed of a superposition of three terms of the form of Equation 2: one with current of -2I positioned at (0,0) and two with current +I positioned at (0,b) and (0,-b), where b is the vertical distance between arrays. For the staggered arrays the two arrays with +I current are centered at (d/2,b) and (d/2,-b).

Table 1 lists the parameters used in the model to estimate remediation times and energy costs. Because of the mathematical singularity of the potential function at the electrodes, the voltage at the electrode must be calculated at the radius of the electrode. In this analysis the electrode radius is 7.6 cm. The remediation times for the two electrode geometries are plotted as a function of horizontal spacing in Figure 4. Both geometries converge to the flat plate solution at vanishing electrode spacing. It is clear that for practical electrode spacing, the staggered array provides much faster remediation. This reflects the more uniform distribution of the electric field or current density that is achieved with the staggered array.

TABLE 1. Model Parameters

Parameter	Symbol	Value
Applied voltage	ΔV	150 V
Vertical spacing	b	2 m
Horizontal spacing	d	2 to 10 m
Electrode radius	e	0.076 m
Electrode length	L	30 m
Contaminated area	A	30 m x 30 m
Ion mobility (Chromate)	m	$5.87 \cdot 10^{-8}\ m^2 V^{-1} s^{-1}$
Chromate concentration	c	100 ppm
Soil background conductivity	σ	$0.042\ \Omega^{-1} m^{-1}$
Soil porosity	n	0.45
Soil tortuosity	τ	1.22
Soil heat capacity	C	$3.5\ kJ\ L^{-1} K^{-1}$

Modeling Calculations. A spreadsheet (Excel™ for the Macintosh™) was developed which incorporated the following:

• A fitted expression for the remediation time vs. electrode spacing for the staggered array geometry, as plotted in Figure 4.

• The analytical expression for the current as a function of the applied potential difference between the anodes and cathodes, the soil resistivity, and the system geometry.

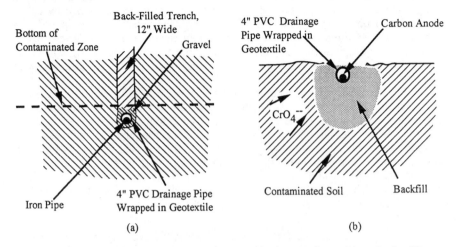

Figure 3. Vertical cross-sections of electrode installations; (a) cathode, (b) anode.

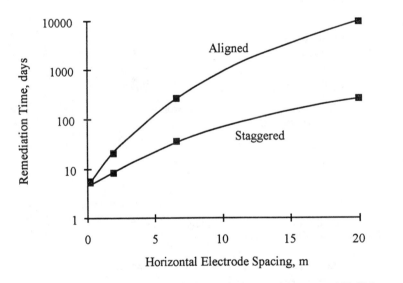

Figure 4. Remediation times for aligned and staggered arrays at 150 V for 2-meter deep contamination.

In this paper the effects of number of electrodes, applied voltage, depth of contamination, and chromate concentration on remediation costs are examined. For a given set of input parameters the spreadsheet calculates the following operating characteristics:

- System current
- Remediation time
- Electrical power
- Remediation energy
- Rates of generation of acid at the anode and base at the cathode
- Water flow rate required to maintain anolyte and catholyte pH between 11 and 11.5
- Adiabatic temperature rise assuming all energy was dissipated in the two-meter deep contaminated soil
- Amount of ion exchange resin required to capture the anions purged from the contaminated soil

Cost Analysis. The data shown in Tables 1 and 2 and the concepts for design and installation of an electrokinetic remediation system were supplied to Paul Amman of the Brattle Group for a detailed cost analysis. The Brattle Group completed the conceptual design including details such as sizes and types of steel and PVC pipe; trailers for electrical equipment, water treatment, and office/lab; and methods of assembly and installation of anodes and cathodes.

Their cost estimate included mobilization to the site and demobilization, purchase and installation of all equipment and instrumentation, engineering and supervision, home office administration, maintenance, utilities, sampling and analysis, insurance, process materials and supplies, waste disposal, and contingencies. Costs that were not included were royalty fees and profit to the remediation vendor. Project cost analysis was done using the number of pairs of electrodes as a variable parameter. Site characterization was not included because it is common to most remediation methods.

Results and Discussion

Modeling Calculations. The results of the spreadsheet calculations for the parameters listed in Table 1 are shown in Table 2 and in Figure 5. Note the large tradeoff in power (and current) versus remediation time. The total energy required, however, varies less than a factor of two if the solution for two pairs of electrodes is excluded. The effects of varying the applied voltage are illustrated in Figure 6. These calculations are for five pairs of electrodes. As expected, remediation time is inversely proportional to the applied voltage, and power is proportional to the square of the voltage.

The energy required for 10 electrode pairs, 62 MWh, is close to the value one would calculate for one-dimensional geometry (60 MWh). Of this energy, about 35 percent is associated with the transport of chromate, and the rest is dissipated by transport of background ions and soil conductivity. The calculated average adiabatic temperature rise in the soil can only provide a rough guideline to the maximum acceptable energy input. The actual rise in soil temperature will need to be modeled or measured in field trials because heat generation will be higher in regions of high current density near electrodes, some heat will be removed in the recirculating solutions, and heat will be lost through the boundaries of the system. In addition,

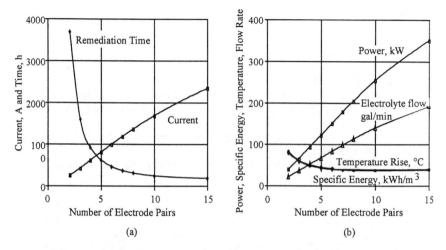

Figure 5. Effect of number of electrodes (for 150 V, 2-meter deep contamination, 100 ppm Cr); (a) on remediation time and current, (b) on power, electrolyte flow, temperature rise, and specific energy.

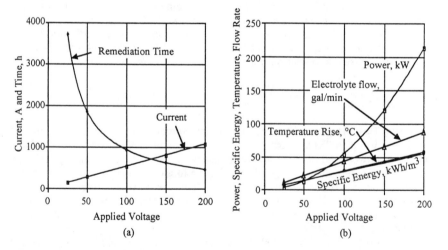

Figure 6. Effect of applied voltage (for 5 electrode pairs, 2-meter deep contamination, 100 ppm Cr); (a) on remediation time and current, (b) on power, electrolyte flow, temperature rise, and specific energy.

heating would increase the mobility of the chromate and other ions and would reduce the total energy input if the voltage were reduced in order to maintain constant current and remediation time. The calculated temperature rise in the soil is proportional to the voltage and will probably be more of a limiting factor in the speed of remediation than power cost. For instance if the voltage were increased to 300 V, the calculated temperature rise for five pairs of electrodes would be 86°C, suggesting that steam would likely be formed near the electrodes, while the energy of 83 kWh/m^3 of soil would cost a reasonable $7/m^3.

Table 2
Spreadsheet Calculations for the Set of Parameters in Table 1

No. of Electrode Pairs**	Horiz. Spacing, meters	Remed. Time, hr	Electr. Power, kW	Remed. Energy, kWh/cu. m	Adiabatic Temp. rise °C	Recirc. Flow Rate, gpm§	Ion Exch. Capacity, equivalents
2	15	3671	38	78	81	20.8	34755
3	10	1593	64	57	59	34.9	25322
4	7.5	916	92	47	49	50.2	20903
5	6	617	120	41	43	65.8	18491
6	5	461	149	38	40	81.4	17081
7	4.29	370	177	36	38	96.6	16244
8	3.75	312	203	35	37	111.1	15766
10	3	246	252	35	36	137.9	15454

** No. of cathodes = no. of pairs + 1
§ To maintain pH between 11 and 11.5

Varying the depth of the contaminated bed for the case of five electrode pairs has the effects shown in Figure 7. The remediation time increases approximately as the square of the depth for thicker beds but does not go to zero at zero depth because the anode to cathode spacing approaches a lower limit of 3 meters. Over a reasonable range of bed depths from 2 to 8 meters, the total energy input ranges only from about 35 to 42 kWh/m^3. Thus the major impact on cost from increasing the bed depth would be from the increased operating costs. The energy requirements predicted by this model are in agreement with reported one-dimensional experimental values on lead and zinc removal[3,4] of about 40 kWh/m^3. The higher energy requirements shown in field tests[7] may be attributable to retardation effects and addition of conditioning agents at the electrodes.

Figure 8 show the results of varying the chromate concentration for the case of five electrode pairs. For a fixed voltage the current is proportional the concentration while remediation time remains constant. The calculated temperature rise of over 140°C for 1000 ppm chromium shows that reducing the voltage would be necessary. At 45 volts the remediation time for five electrode pairs would be 2058 hr, the total energy would be 43 kWh/m^3, and the adiabatic temperature rise would be 44°C. The increased remediation cost would be a combination of higher operating cost plus greater water treatment and waste disposal costs.

The system is marginally sensitive to variations in the diameter of the electrodes in the range of 8 to 20 cm. Reducing the diameter from 15.2 cm to 10.2 decreases the

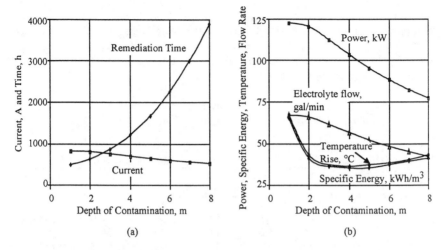

Figure 7. Effect of depth of contamination (for 5 electrode pairs, 150 V, 100 ppm Cr); (a) on remediation time and current, (b) on power, electrolyte flow, temperature rise, and specific energy.

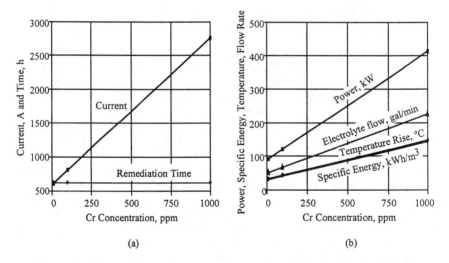

Figure 8. Effect of chromium concentration (for 5 electrode pairs, 150 V, 2-meter deep contamination); (a) on remediation time and current, (b) on power, electrolyte flow, temperature rise, and specific energy.

current and raises the time each by 11% to 12% and does not significantly affect the total energy required.

While the results of this model are consistent with reported experimental results, several assumptions and approximations limit the accuracy of the calculations. Constant and uniform soil conductivity will not be the case in the field because of temperature variations, ion migration, and chemical reactions. The model assumes that chromate removed at the anode is adsorbed on ion exchange resin and that other anions desorbed from the resin will be introduced at the cathode. Therefore the resistivity of the water should be maintained roughly constant. The control of the pH might require the addition of sufficient acid or base to raise the conductivity of the water. This increase could be compensated for by replacing some of the water with fresh water. If the anions removed at the anode were not replaced with anions at the cathode, the soil electrical resistivity would rise. For a given applied voltage, this would result in a reduction in current and a lowering of input power. The chromate ion mobility used here is the handbook value for infinite dilution[8]. In soils at finite concentration the actual mobility may differ because of adsorption, complexation, or concentration effects. End and edge effects were not taken into account for the finite conceptual site, and these could add 20% or more to either remediation time or energy. Other approximations such as soil porosity and tortuosity, background soil resistivity, and soil heat capacity, are within expected ranges, but could vary considerably. If there is retardation of chromate by adsorption, or the conductivity increases because of salt buildup, the energy requirements will increase. For soils of differing conductivities, the energy requirements can be estimated by applying the linear relation between power consumption and soil conductivity. The background soil conductivity chosen in this model is typical for soils, however this value may be low for waste sites. This fact is addressed in the model by adjusting the soil resistivity for the concentration of chromate.

Cost Analysis. The breakdown of the estimated costs for the case of 5 anodes and 6 cathodes, 100 ppm Cr, 150 volts, and 2 meter deep contamination is shown in Table 3. The estimated weight of the 1800 m^3 of soil was 3600 metric tons.

The variations, vs. remediation time, of allocated capital, operating, and total unit costs are shown in Figure 9a. Because of the tradeoff between operating (direct) costs and capital costs, there is a minimum at 35 to 50 days which corresponds to 4 or 5 pairs of electrodes. Energy cost, including a site base load of 6 kilowatts, is a small part of the total as shown in Figure 9b. Electrodes and electrical equipment, parts of which are reusable, represent only 15% to 30% of the unit costs.

Rough estimates of the effects on costs of increasing the chromium concentration or of increasing the depth of the contaminated bed were made using the assumption that direct costs and capital costs shown in Figure 9a would be applicable with slight modifications. For example, for 1000 ppm chromium, the voltage would need to be reduced to about 45 volts to keep the soil temperature at the same level as shown in Table 2. For six pairs of electrodes, the on-site time would then increase to 78 days. The capital cost for six pairs of electrodes would be about $106/ton and the operating cost for 78 days would be about $35/ton. About $8/ton in added ion exchange resin would be needed for a total cost of about $149/ton. When this exercise is done for different numbers of electrode pairs, a cost minimum occurs at six pairs with very little difference in total cost for four to eight pairs. Because complete cost analyses were not done for this case or for the case considered below, there are uncertainties of several dollars per ton. However, the trends in costs should be correct.

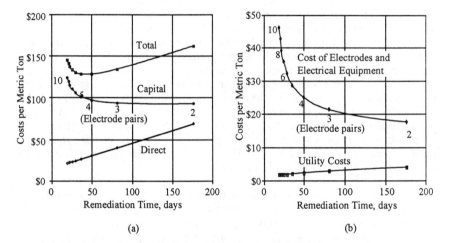

Figure 9. Costs per metric ton of soil; (a) direct, capital and total costs, (b) costs of utilities and of electrodes and electrical equipment.

Table 3
ESTIMATED REMEDIATION COSTS

1.   Fixed capital costs

Anodes (installed)	$16,740
Cathodes (installed)	48,726
Electrical systems	37,582
Electrolyte systems[a]	167,000
Site equipment	102,000
Other costs [b]	100,000
Indirect costs [c]	194,672

Total fixed capital $666,720

Portion of fixed capital costs allocated to the site

Reusable equipment [d]	$4,319
Site specific equipment	360,138

Subtotal $364,457

2.   Operating costs

Process materials[e]	$35,646
Utilities	6,964
Operation	5,399
Maintenance	2,818
Mobiliz./demobiliz.	27,500
Sampling and Analysis	8,227
Indirect costs [f]	8,023
Disposal	1,914

Total direct costs $96,490

3.   Total project costs

Total costs	$460,947
Unit costs per metric ton	$128

[a]   Tanks, pumps, and controllers of ion exchange system
[b]   Instrumentation and control and miscellaneous
[c]   Engineering and supervision, construction management, and contingency
[d]   Depreciation on reusable equipment (electrodes, power supplies, etc.) during the time used on site, based on a five year lifetime. Reusable equipment is approximately 45% of all capital equipment.
[e]   Mostly ion exchange resin
[f]   Insurance, home office administration, travel, per diem expense, auto rental

Increasing the contaminated soil depth to 4 meters would require an increase in voltage to 175 volts to maintain the same energy input per ton of soil and the same temperature rise. The major decrease in unit cost would be due to spreading operating and capital costs over twice the mass of contaminated soil. Offsetting these decreases would be increases in the cost of cathodes (assumed to double because of deeper installation), double the amount of ion exchange resin, and less than $1/m^3$ additional energy cost. Small increases in unit costs from increases in the size of the electrical and water treatment systems were not taken into account. The rough estimate of remediation cost is about $76 per metric ton. The cost optimum occurs at three to five pairs of electrodes with on-site times between 55 and 96 days. Significantly thicker contaminated beds would require different methods of cathode installation, and this cost analysis would not apply.

Conclusions

This analysis of electrokinetic remediation of soluble metal ions is useful, within the limitations of the assumptions and simplifications, for estimating approximate remediation costs, for examining the sensitivity of costs to system design parameters, and for guiding the approach to system design. Some specific conclusions are listed below:

- A staggered array of electrodes is preferable to an aligned array because energy costs and remediation times are smaller.
- For the staggered array, current (or power) and remediation time vary strongly in opposite directions as the electrode spacing changes. Total energy, the product of power times time, varies much less.
- The major tradeoffs are between electrode spacing and time, and between applied voltage (or power input) and time.
- Temperature rise in the soil, and not energy cost, will be the controlling factor in choosing system voltage for this geometry. Energy cost is a small fraction of the total cost.
- Remediation times for mobile metal ions are in a reasonable range of a few weeks to a few months.
- Estimated total remediation cost is in the range of $130 to $140 per metric ton of soil for 100 ppm chromium in the ground water in a 30 m x 30 m x 2 m bed of soil.
- Estimated remediation costs increase only to about $150 per metric ton if the chromium contamination is increased to 1000 ppm.
- If the depth of contamination is increased to 4 meters, estimated total remediation costs decrease significantly, possibly to as low as $76 per metric ton.

Acknowledgments

The work of Paul Amman and his co-workers at the Brattle Group was a critical part of this analysis. They not only provided the careful cost analysis, but took some of our design concepts and developed them into useful engineering plans.

References
1. Lageman, R.; Pool, W.; Seffinga, G., *Chemistry & Industry,* **1989**, 585.
2. Shapiro, A. P.; Probstein, R. F., *Environmental Science and Technology,* **1993**, 27(2), 283.
3. Probstein, R. F.; Hicks, R. E., *Science,* **1993**, 260, 498.
4. Hamed, J.; Acar, Y. A.; Gale, R. J., *Journal of Geotechnical Engineering,* **1991**, 117(2), 241.
5. Shapiro, A. P.; Renaud, P. C.; Probstein, R. F., *PhysicoChemical Hydrodynamics,* **1989**. 11(5/6), 785.
6. Bear, J., *Dynamics of Fluids in Porous Media,* Dover Publications, Inc.: New York, 1972; p. 322.
7. Lageman, R., *Environ. Sci. Technol.,* **1993**, 27(13), 2648.
8. Weast, R. C., ed., *CRD Handbook of Chemistry and Physics,* 65th Edition, CRC Press, Inc., 1984; p. D-172.

RECEIVED March 14, 1995

Chapter 5

Numerical Simulation of Electrokinetic Phenomena

Eric R. Lindgren[1], R. R. Rao[2], and B. A. Finlayson[3]

[1]Sandia National Laboratories, P.O. Box 5800, MS 0719
and [2]MS 0827, Albuquerque, NM 87185
[3]Department of Chemical Engineering, BF-10, University of Washington, Seattle, WA 98195

In this paper, we outline our approach to modeling electrokinetic phenomena in unsaturated porous media. A short summary of previous electrokinetic modeling efforts for other media is also given. One issue of fundamental importance to electrokinetic modeling is whether a formulation exhibits electrical neutrality at all times. A proof demonstrating that the equations used ensure electroneutrality is presented. In addition, some preliminary modeling results are presented for a system of all strong electrolytes and the ground work is laid out for solving more complex problems needed for environmental remediation work.

Recently, researchers at Sandia National Laboratories have been conducting experiments to demonstrate the feasibility of using electrokinetic remediation in unsaturated soils (*1-3*). Anionic dye has been shown to move in unsaturated soils with moisture contents as low as 4 wt%. Figure 1 illustrates a typical experimental set-up in a somewhat idealized fashion. A test cell is packed with a soil moistened with a solution of calcium chloride to a desired moisture content. The center strip of soil is removed and replaced with soil having the same moisture and calcium chloride content but contaminated with the disodium salt of a food dye that behaves as a strong electrolyte. The pore water concentration of the dye is four times the calcium chloride concentration. The pH is near neutral so the H^+ and OH^- concentrations are negligible compared to Ca^{2+} and Cl^-. At both ends of the cell, the soil contacts porous ceramic casings that contain graphite electrodes in an electrolyte solution held under negative pressure or tension. The tensions of the electrolyte solutions are set to match the soil tension at the desired moisture content. The electrolyte solution is continuously circulated past the electrodes to remove contaminants and condition electrode reactions. When constant direct current is passed through the cell, the initial voltage profile is characterized by three straight lines with sharp slope transitions at the dye interfaces where the total ionic concentration changes abruptly. The chemistry of the electrolyte solutions is such that the hydrogen ions produced at the anode instantly react and calcium ions migrate toward the cathode; the hydroxyl ions produced at the cathode react and chloride ions migrate toward the anode. The electrolyte concentrations and flow rates can be controlled such that the calcium and chloride concentrations remain

0097–6156/95/0607–0048$12.00/0

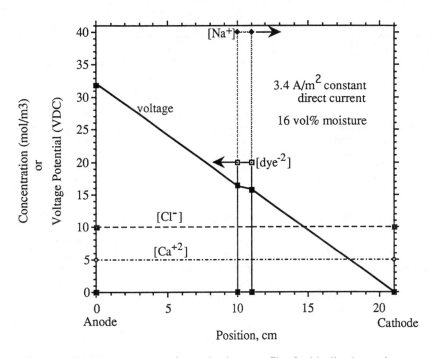

Figure 1. Initial ion concentration and voltage profiles for idealized experiment.

constant near the electrode casings. Electroosmosis causes moisture to move slowly from the anode to the cathode. Since water is removed at the cathode and replenished at the anode, electroosmosis causes little change in the moisture distribution across the test cell. The dye anions begin to migrate toward the anode and the sodium cations begin migrating toward the cathode. How can these ions separate without violating electroneutrality?

The answer is not intuitive. A thorough understanding of the physics behind the contaminant migration phenomena is required to predict such behavior and optimize the process for environmental remediation. The problem is highly coupled and nonlinear because the migration rate of any ion depends on the local potential gradient, which is dependent on the current density and the concentration of all the ions at that point. The situation in unsaturated soil is further complicated by the dependence of the pore water current density on the soil moisture content that may change with time and position. Because of the complexity of the problem, numerical methods are required to solve the equations. This paper summarizes the equations used in previous modeling efforts, outlines how we approach the problem, presents a proof that demonstrates that the equations used ensure electroneutrality, and presents some preliminary results.

Background

Electrokinetic phenomena have been used for many years to separate and analyze mixtures of complex organic compounds such as proteins in solution. A number of different electrophoretic separation techniques have been successfully modeled (4-9). The electrokinetic phenomena utilized in electrophoretic separation techniques are the same as those used in the electrokinetic remediation process under development at Sandia. Differences arise in the initial and boundary conditions and in the material containing the electrolyte solution (capillary tubes versus soil). Shapiro (10) extended the model of Saville and Palusinski (7) for use with electrokinetic remediation processes in saturated soil by introducing tortuosity and porosity, and allowing electroosmosis.

All of the above models determine the evolution of the electric field and concentration of each species in one dimension. Balances on mass and charge are developed into a system of convective diffusion equations with reaction that are solved subject to a number of constraints. Following the notation of Shapiro et al. (11) and Probstein (12) a mass balance inside a capillary tube or soil pore yields:

$$\frac{\partial c_i}{\partial t} = -\frac{\partial}{\partial x} j_i^* + R_i \tag{1}$$

where c_i is the concentration of species i (mol/m^3), t is time (s), and R_i is the molar rate of production due to chemical reaction (mol/m^3 s). For isothermal conditions, the molar flux, j_i^*(mol/m^2 s), can be due to diffusion, ionic migration in an electric field and convection (hydraulic and electroosmotic):

$$j_i^* = -F\upsilon_i z_i c_i \frac{\partial \phi}{\partial x} - D_i \frac{\partial c_i}{\partial x} + c_i u \tag{2}$$

where υ_i is the absolute mobility of species i (mol m/N s), F is Faraday's constant (96,485 C/mol), z_i is the charge number of species i, ϕ is the potential (V), D_i is the diffusivity of species i (m^2/s), and u is the mass average velocity of the fluid (m/s), which is the sum of hydraulically and electroosmotically driven flows.

All of the above mentioned electrophoretic separation models assume no convection. The model of Shapiro et al. (*11*) includes electroosmotic flows, which depends on the potential gradient, but assumes no hydraulic flow.

A mass balance on each species alone is insufficient to solve the problem. A relationship between the electric potential and the species concentration is needed to obtain a solution. A balance on electric charge yields:

$$\frac{\partial \rho_e}{\partial t} = -\frac{\partial}{\partial x} F\Sigma \, z_i \, j_i^*$$

(3)

where ρ_e is the electric charge density in (C/m^3) and $F\Sigma \, z_i j_i^*$ is the charge flux (C/s m^2). In solutions, the charge density is defined as

$$\rho_e = F\Sigma \, z_i \, c_i$$

(4)

Substitution of this identity into equation 3 yields

$$\Sigma \, z_i \frac{\partial c_i}{\partial t} = -\frac{\partial}{\partial x} \Sigma \, z_i \, j_i^*$$

(5)

The assumption is typically made that the solution is electrically neutral at all points meaning the charge density is zero.

$$\Sigma \, z_i \, c_i = 0$$

(6)

Based on this identity, equation 5 reduces to

$$\frac{\partial}{\partial x} F\Sigma \, z_i \, j_i^* = 0$$

(7)

Integration of equation 7 yields the current density equation:

$$J = F\Sigma \, z_i \, j_i^*$$

(8)

where the constant of integration, J, is the current density (A/m^2 or C/s m^2). Thus, the current equation embodies the assumption of electroneutrality.

All that remains is to define the reaction term in equation 1. For strong electrolytes, the reaction term is zero. For most other species, this term poses a problem because most pertinent reactions are fast equilibrium reactions that render the problem impossibly stiff to solve numerically. As discussed by Saville and Palusinski (*7*) and Shapiro (*10*), the sum of the reaction terms for each form of a weak electrolyte (associated or dissociated) is zero. Thus, the reaction term for a weak electrolyte can be eliminated from the system of equations by a linear combination of the corresponding mass balance equations (equation 1) and the equilibrium relationship. This combination requires the assumption that the diffusivity of each form of the electrolyte is the same, which is true only at infinite dilution (*13*). Because the diffusivity of H^+ and OH^- differ significantly, this procedure will not work to eliminate their reaction terms. To eliminate these reaction terms, the water equilibrium relationship and the electroneutrality relationship (equation 6) are used to solve for the H^+ and OH^- concentrations. This procedure is justified provided the system of equations to be solved ensures electroneutrality at all points. Any numerical error resulting in deviation from electroneutrality accumulates in the H^+ and OH^- ion concentrations calculated by this method.

Only in the paper describing the model of Dose and Guiochon (9) is the inherent nature of their formulation to maintain electroneutrality discussed and a formulation specific proof offered. In this particular model, only strong electrolytes were considered and so the issue of the H^+ and OH^- reaction rates is not discussed. All of the other models do consider H^+ and OH^- ions and use electroneutrality and equilibrium constraints to determine the concentrations.

Electroneutrality

It is a simple matter to show that the formulation given above inherently ensures electroneutrality at all points and for all times. First, we multiply equation 1 by z_i and sum over i

$$\Sigma z_i \frac{\partial c_i}{\partial t} = -\Sigma \frac{\partial}{\partial x} (z_i j_i^*) + \Sigma z_i R_i \tag{9}$$

The last term on the right-hand-side is zero because there is no net charge produced by a chemical reaction and we are left with the current balance as shown in equation 5

$$\Sigma z_i \frac{\partial c_i}{\partial t} = -\frac{\partial}{\partial x} \Sigma z_i j_i^* \tag{5}$$

As discussed earlier, electroneutrality must be assumed to obtain the current density equation which gives

$$\Sigma z_i \frac{\partial c_i}{\partial t} = \frac{\partial}{\partial t} \Sigma z_i c_i = 0 \tag{10}$$

If the total charge density is zero at time zero for any position then it is zero for all time.

This means that any general formulation that combines a system of convective-diffusion equations with the current density equation should ensure electroneutrality. This is true of all the models discussed above. However, it would be prudent to check any numerical implementations on a simplified test case (i.e., with strong electrolytes), such as shown in Figure 1, where any problems maintaining electroneutrality will be immediately evident. Only after numerical electroneutrality has been demonstrated should the electroneutrality constraint be used to calculate other ion concentrations, especially H^+ and OH^-. Any numerical errors maintaining electroneutrality will accumulate in the H^+ and OH^- concentrations and because of their high mobility, will feed back into the potential gradient calculation and effect the movement of all ions.

Model Formulation

We follow the work of Shapiro (10) to transform the equations presented above from a single capillary tube model to a general porous media model. The transformation from capillary coordinates, x, to porous medium coordinates, z, is

$$\frac{dz}{dx} = \frac{1}{\tau} \text{ or } \frac{\partial}{\partial x} = \frac{\partial}{\partial z} \frac{dz}{dx} = \frac{1}{\tau} \frac{\partial}{\partial z} \tag{11}$$

Pore tortuosity, τ, accounts for the actual pore path length being longer than the path through a straight pore parallel to the voltage gradient.

To convert from the applied current, I (A or C/s), to the pore water current density, J, we divide the applied current by the cross sectional area available for current transport, *e.g.*, the actual area that is continuously wet. For unsaturated porous media, this is described by the moisture content, θ, multiplied by the test cell cross sectional area.

$$J = I/(\theta A) \tag{12}$$

Rather than work with the convective diffusion equation in its divergence form given by equations 1 and 2, our approach is to substitute equation 2 into equation 1 and expand the derivatives. We have also inserted the conversion from capillary tube to porous media coordinates.

$$\frac{\partial c_i}{\partial t} = \frac{F\upsilon_i z_i}{\tau^2}\left(c_i \frac{\partial^2 \phi}{\partial z^2} + \frac{\partial c_i}{\partial z}\frac{\partial \phi}{\partial z}\right) + \frac{D_i}{\tau^2}\frac{\partial^2 c_i}{\partial z^2} + R_i \tag{13}$$

Here, the simplification of no convection has been made. When equation 2 is substituted into equation 8 and transformed by equation 11, we can solve for the potential gradient:

$$\frac{\partial \phi}{\partial z} = -\frac{1}{\sigma}\left(\tau J + F\sum_{j=1}^{n} z_j D_j \frac{\partial c_j}{\partial z}\right) \tag{14}$$

where the conductivity, σ, is defined (for infinite dilution) by:

$$\sigma = F^2 \sum z_i^2 \upsilon_i c_i \tag{15}$$

Differentiating equation 14 yields:

$$\frac{\partial^2 \phi}{\partial z^2} = \frac{1}{\sigma^2}\frac{\partial \sigma}{\partial z}\left(\tau J + F\sum_{j=1}^{n} z_j D_j \frac{\partial c_j}{\partial z}\right) - \frac{1}{\sigma}\left(F\sum_{j=1}^{n} z_j D_j \frac{\partial^2 c_j}{\partial z^2}\right) \tag{16}$$

where the electroneutrality assumption (equations 6 and 7) is used to set $\frac{\partial J}{\partial z} = 0$.

Equations 14 and 16 are substituted into equation 13 to eliminate the potential gradient dependence from the system of convective diffusion equations giving:

$$\frac{\partial c_i}{\partial t} = \frac{F\upsilon_i z_i}{\tau^2}\left(\frac{1}{\sigma^2}\frac{\partial \sigma}{\partial z}\left(\tau J + F\sum_{j=1}^{n} z_j D_j \frac{\partial c_j}{\partial z}\right) - \frac{1}{\sigma}\left(F\sum_{j=1}^{n} z_j D_j \frac{\partial^2 c_j}{\partial z^2}\right)\right) -$$
$$\frac{F\upsilon_i z_i}{\tau^2}\frac{\partial c_i}{\partial z}\frac{1}{\sigma}\left(\tau J + F\sum_{j=1}^{n} z_j D_j \frac{\partial c_j}{\partial z}\right) + \frac{D_i}{\tau^2}\frac{\partial^2 c_i}{\partial z^2} + R \tag{17}$$

It can be shown that this formulation ensures electroneutrality as expected from the more general proof given earlier.

In order to include H^+ and OH^- ions, the electroneutrality and water equilibrium constraints will be used.

$$\sum z_i c_i = 0 \qquad (6)$$

$$c_H\, c_{OH} = K_w \qquad (18)$$

It is desirable to maintain the set of equations to be solved as a set of partial differential equations (PDE) and avoid mixed differential/algebraic equation sets, which can be very difficult to solve. We therefore take the time derivative of equations 6 and 18

$$\sum z_i \frac{\partial c_i}{\partial t} = 0 \qquad (19)$$

$$c_H \frac{\partial c_{OH}}{\partial t} + c_{OH} \frac{\partial c_H}{\partial t} = 0 \qquad (20)$$

These equations can be solved simultaneously to give:

$$\frac{\partial c_H}{\partial t} = -\frac{\sum\limits_{i=3}^{n} z_i \frac{\partial c_i}{\partial t}}{(1 + \frac{c_{OH}}{c_H})} \qquad (21)$$

$$\frac{\partial c_{OH}}{\partial t} = -\frac{c_{OH}}{c_H} \frac{\partial c_H}{\partial t} \qquad (22)$$

Substitution of these two equations for the corresponding convective diffusion equations should impose the desired constraints without disruption of the PDE solution scheme. The computational effort is of course increased by including the H^+ and OH^- ions.

Poisson's Equation & the Electroneutrality Paradox

Comparison of the current density equations solved for the derivatives of the potential with Poisson's equations suggests a paradox. By differentiating equation 14 we can obtain the Laplacian of the electric potential, *i.e.*, equation 16

$$\frac{\partial^2 \phi}{\partial z^2} = \frac{1}{\sigma^2} \frac{\partial \sigma}{\partial z} (\tau J + F \sum_{j=1}^{n} z_j\, D_j\, \frac{\partial c_j}{\partial z}) - \frac{1}{\sigma} (F \sum_{j=1}^{n} z_j\, D_j\, \frac{\partial^2 c_j}{\partial z^2}) \qquad (16)$$

However, for electrolyte solutions in porous media, Poisson's equation may be used to obtain another expression for the Laplacian of the electric potential:

$$\frac{\partial^2 \phi}{\partial z^2} = \frac{\tau^2 F}{\varepsilon} \sum z_i\, c_i \qquad (23)$$

where ε is the permittivity of water (6.9×10^{-10} $C^2 m^2/N$ or $C/V\, m$) and F is Faraday's constant (96,485 C/mol). Poisson's equation is derived from Maxwell's equations

(13) and in principle must be satisfied at all points and times. The large value of $\tau^2 F/\varepsilon$, *e.g.*, 10^{14} V-m/mol, means that negligible deviation from electroneutrality results in significant non-zero values of the electric potential Laplacian *(14)*.

Equation 16 seems to conflict with Poisson's equation. The electroneutrality assumption embodied in the current equation demands, through Poisson's equation, that the potential gradient be constant. Since equation 16 is used to calculate changes in the potential gradient, there appears to be a paradox. An explanation can be found in the physical interpretation of these two equations. Poisson's equation depends on the actual, although very small, charge separations present in the system, essentially at the atomic level. The current equation, on the other hand, depends on continuum level diffusion processes that produce the very small charge separations in the system.

Consider the spike experiment depicted in Figure 1 when no current is applied. The dye and the sodium ions will both diffuse away from the initial boundary. Since the sodium ions have a greater mobility than the dye ions, the sodium ions will diffuse ahead of the dye ions separating charge. As the charges are separated, a local electric field will develop which retards the sodium ion mobility and augments the dye ion mobility. Equilibrium will quickly be established and the dye and the sodium ions will diffuse at the same rate, however the sodium ions will be slightly ahead of associated dye ions. Whenever there are concentration gradients and diffusion is occurring, electroneutrality will be violated and the Laplacian of the electric potential will be non-zero. A measurable potential gradient called the diffusion potential will develop across the diffusion front *(13, 14)*. This potential gradient could, in principle, be determined using Poisson's equation, however, all the charge separations would have to be known on a microscopic scale that would be nearly impossible to calculate numerically. In practice, the diffusion potential is calculated using the current equation with the current set to zero *(13, 14)*. Thus, even though the current equation requires the electroneutrality assumption, it can be used to accurately calculate the potential field that develops as a result of small charge separations created by the diffusion process. Even though deviations from electroneutrality can cause the Laplacian of the electric potential to be non-zero, the assumption of electroneutrality is consistent with the assumption of solution continuum. The current equation should be as accurate as the diffusivities on which it is based. However, at concentrations higher than 10^{-3} M, concentration dependence of ion diffusivity and mobility should be considered *(13)*.

Numerical Method and Results

A numerical implementation has been made for the strong electrolyte case ($R_i = 0$) with no external convection such as electroosmotic or hydraulic flow. Here we use an arbitrarily chosen tortuosity of 1.5 and solve equation 17 for each ionic species of interest for the case depicted in Figure 1. The boundary conditions used are somewhat arbitrarily set at $dc_i/dz = 0$ at the electrodes.

Because of the nonlinear nature of the problem and the strong coupling between all the species concentrations caused by their dependence on the potential gradient, these equations are challenging to integrate and special techniques must be used. The numerical method used to integrate the partial differential equations is a method of lines. Here, a finite difference technique is used to discretize the spatial dimension and a stiff equation solver with variable time step size is used to integrate the resulting coupled ordinary differential equation set in time. The finite difference method used is particularly suited to problems with sharp fronts since it has an adaptive mesh capability that uses a finer mesh in the regions where the solution is changing rapidly. It can solve the problem very efficiently since mesh is added or destroyed depending on the gradient of local solution *(15)*. In principle, this method can be used to solve problems for any number of species, but of course the number of

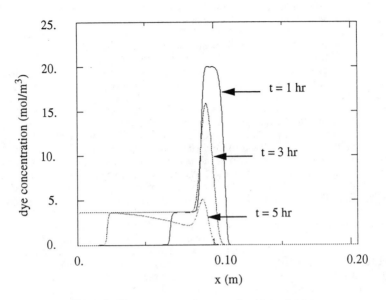

Figure 2. Dye concentration as a function of time.

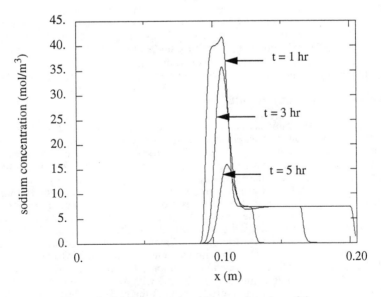

Figure 3. Sodium concentration as a function of time.

equations to be solved and thus the computational effort increases with the number of species considered.

Figures 2 through 5 show the species migration as a function of time for the dye, sodium, chloride, and calcium ions. The behavior of the ions is affected by three important forces: diffusion causes the ions to move in both directions, the imposed electric field causes the negatively charged ions to move toward the anode and the positively charged ions to the cathode, and finally the local field created by small deviations from electroneutrality influences ion motion. The dye and sodium ion concentration began as spikes in the center of the domain at the beginning of the experiment. Over time the ions move toward the electrode creating an almost bimodal distribution. The concentration near the original spike location remains high, but gets thinner, while a lower concentration front advances toward the electrode. Over time the spike region concentration becomes lower while the concentration associated with the advancing front stays constant.

For the ions that began the experiment with a constant concentration distribution, chloride and calcium, we observe different behavior. We see a depletion front between the spike region and the attracting electrode (Cl^- and the anode; Ca^{2+} and the cathode) and a build up of concentration in the spike zone. This effect answers the question asked in the introduction as to how the dye and sodium can separate to opposite electrodes and still maintain electroneutrality. This separation is accomplished by other ions moving into the center region from which the sodium and dye ions departed thus keeping the system electrically neutral. This view point is dramatically demonstrated in Figure 6, which shows all ion concentrations after 5 hours. We can see that where we previously had a sodium and dye spike, after 5 hours of electromigration, calcium and chloride peaks have formed and much of the dye and sodium has left the domain through the electrodes. If we look at the potential gradient, Figure 7, we can see that this field approaches zero in the center of the domain and is almost a constant value in the other regions where the concentration gradients are not as great. Note, the bulk system has stayed electrically neutral to 10^{-8} mol/m^3 or better at each grid point in the domain, even after 5 hours.

It is interesting to compare the Laplacian of the voltage calculated from differentiating the current equation and from Poisson's equation (Figures 8 and 9, respectively). From the current equation we get a field that is smooth and has large peaks where the ionic strength changes dramatically. If we use Poisson's equation to calculate this field, we get qualitatively similar behavior but the results are not smooth and fluctuate rapidly from large positive numbers to large negative numbers over small spatial scales. In addition, the peak values of the Laplacian of the potential are 5 orders of magnitude higher if Poisson's equation is used. However, the peaks in Figure 9 at the boundaries are an artifact of the unrealistic boundary conditions that have been used: these boundary conditions have not been formulated to be electrically neutral. Away from the boundaries the Laplacian of the voltage calculated from Poisson's is closer to the values calculated from the current equation. However, the solution still oscillates wildly over small changes in x, demonstrating quantitatively why using Poisson's equation for the second derivative of the potential makes the resulting equation set extremely stiff and impossible to integrate.

Conclusions

Simultaneous solution of the coupled species mass balances with the current equation substituted in for the potential gradients has been shown to ensure electroneutrality. Therefore, all of the models reviewed from the literature and the model described here should ensure electroneutrality. However, before using the electroneutrality constraint to eliminate a convective-diffusion equation, the model should be tested on a problem (such as the one presented here) where deviations from electroneutrality will be

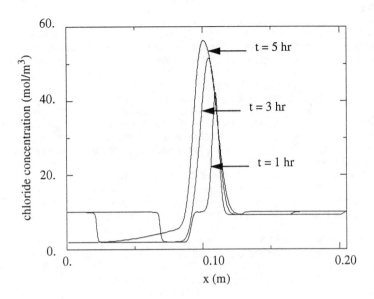

Figure 4. Chloride concentration as a function of time.

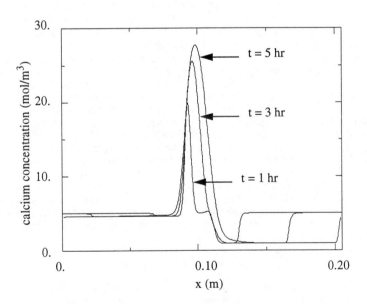

Figure 5. Calcium concentration as a function of time.

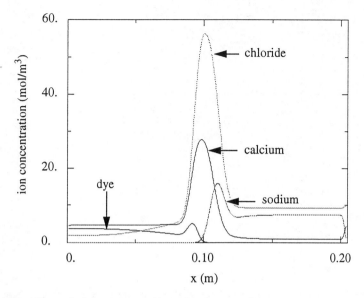

Figure 6. Ion concentrations after 5 hours of constant current.

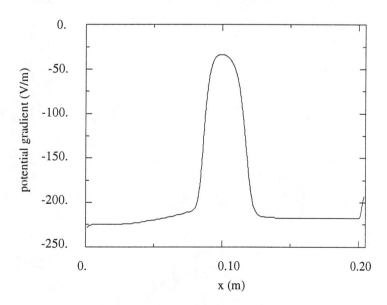

Figure 7. Voltage gradient after 5 hours of constant current.

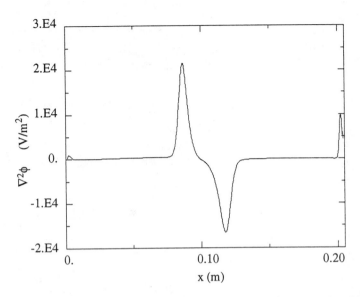

Figure 8. Laplacian of voltage calculated from the current equation after 5 hours of constant current.

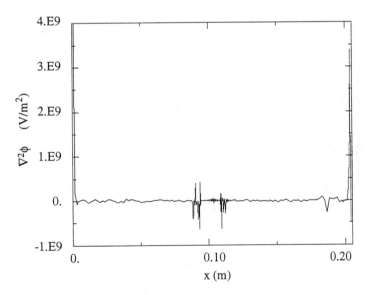

Figure 9. Laplacian of voltage calculated from Poisson's equation after 5 hours of constant current.

immediately evident. Any numerical errors causing deviations in electroneutrality will accumulate in the species concentration calculated using the electroneutrality constraint, which along with water equilibrium is typically used to calculate the hydrogen and hydroxyl ion concentration. Numerical errors in calculating the concentration of these ions will be magnified because the higher mobility of these ions results in a greater influence on the solution conductivity which in turn has a large influence on the potential gradient, the driving force for ion migration.

The electroneutrality assumption applied to the current equation appears to be inconsistent with Poisson's equation, which must hold at all points and times. When applied to solutions, Poisson's equation requires consideration at atomic levels whereas the current equation is based on diffusion in a continuum. Thus, both equations are valid, the electroneutrality assumption in solutions is as valid as the assumption of solution continuum, and the current equation should be as accurate as the diffusivities on which it is based.

Our preliminary modeling results show that complete dye and sodium removal should be possible from a contaminated soil sample. This preliminary work has been carried out to obtain a working code that can later be modified to include more complicated physics. For future work, we plan on extending our model by adding hydraulic flows and electroosmosis, using more realistic boundary conditions, and modeling problems that include acid/base chemistry. This will be the first modeling work done on electrokinetic phenomena in unsaturated systems with hydraulic flow.

Acknowledgments

This work was performed at Sandia National Laboratories which is operated for the U.S. Department of Energy under Contract No. DE-AC04-94AL85000 and was funded by the Office of Technology Development, within the Department of Energy's Office of Environmental Restoration and Waste Management, under the In Situ Remediation Technology Development Integrated Program and the Mixed Waste Landfill Integrated Demonstration.

The authors would like to thank R. J. Gross (Sandia National Laboratories, Department 1512) for his help with the implementation of the electrokinetic equations into an existing program designed to solve stiff equations with sharp fronts. His help greatly simplified an otherwise arduous task.

Literature Cited

1. Lindgren, E.R.; Kozak, M.W.; Mattson, E.D. "Electrokinetic Remediation of Contaminated Soils", Proceedings of the ER'91 Conference at Pasco, WA, Sept 1991, pp. 151.
2. Lindgren, E.R.; Kozak, M.W.; Mattson, E.D. "Electrokinetic Remediation of Contaminated Soils: An Update", Proceedings of the Waste Management 92 Conference at Tucson, AZ, March 1992, pp. 1309.
3. Lindgren, E.R.; Kozak, M.W.; Mattson, E.D. In *Emerging Technologies in Hazardous Waste Management IV* ; Tedder, D.W.; Pohland, F.G., Eds.; ACS Symposium Series 554; ACS: Washington, DC 1994, pp 33-50..
4. Palusinski, O.A.; Allgyer, T.T.; Mosher, R.A.; Bier, M. *Biophys. Chem.* **1981**,*13*, 193.
5. Palusinski, O.A.; Bier, M.; Saville, D.A. *Biophys. Chem.* **1981**,*14*, 389.
6. Bier, M.; Palusinski, O.A.; Mosher, R.A.; Saville, D.A.*Science*, **1983**, *219*, 1281.
7. Saville, D.A.; Palusinski, O.A. *AIChE J.*, **1986**,*32*, 207.

8. Palusinski, O.A.; Graham, A.; Mosher, R.A.; Bier, M.; Saville, D.A. *AIChE J.*, **1986**, *32*, 215.
9. Dose, E.V.; Guiochon, G.A. *Anal. Chem.*, **1991**, *63*, 1063.
10. Shapiro, A.P., "Electroosmotic Purging of Contaminants from Saturated Soils", Ph.D. Thesis, Massachusetts Institute of Technology, 1990.
11. Shapiro, A.P.; Renaud, P.C.; Probstein, R.F. *J. PhysicoChemical Hydro.*, **1989**, *11*, 785.
12. Probstein, R.F., *Physicochemical Hydrodynamics. An Introduction*; Butterworth: Stoneham, MA, 1989.
13. Bockris, J. O'M; Reddy, A.K.N. *Modern Electrochemistry, Volume 1*; Plenum Press, New York, NY, 1970.
14. Newman, J.S. *Electrochemical Systems*; Second Edition; Prentice-Hall: Englewood Cliffs, NJ, 1991.
15. Gross, R.J.; Baer, M.R.; Hobbs, M.L. "XCHEM 1-D: A Heat Transfer Chemical Kinetics Computer Program for Multilayered Reactive Materials", SAND93-1603, Sandia National Laboratories, Albuquerque, NM, 1993.

RECEIVED March 14, 1995

ELEMENT RECOVERY AND RECYCLE

Chapter 6

Treatment of an Anionic Metal by Adsorption on Iron Oxides

E. Lee Bagby and Charles M. West

Resource Consultants, Inc., 320 Southgate Court, P.O. Box 1848, Brentwood, TN 37024–1848

Adsorption of antimony onto amorphous iron hydroxide (ferrihydrite) as a treatment method for industrial wastewater was evaluated. The effects of pH, iron hydroxide to antimony ratio, temperature, time and interference due to sulfate were investigated. Adsorption was found to occur within a narrow pH range at a reasonable antimony to iron hydroxide ratio. Elevated temperature inhibited antimony adsorption. Sulfate also inhibits adsorption but can be overcome with additional iron hydroxide even at relatively high (1.2 molar) sulfate concentration. The reaction was rapid, reaching maximum adsorption within 15 minutes. This treatment has been successfully applied an industrial wastewater.

Antimony is used in the manufacture of lead-acid batteries. The hardness of lead is increased by the addition of 1 to 10 percent antimony. Lead-acid batteries are intensively recycled, and the antimony in these batteries is introduced into the secondary lead smelters.

The soluble antimony content in the sulfuric acid electrolyte of lead acid batteries is relatively modest. Larger quantities of antimony are leached into water due to the use of sodium carbonate to neutralize sulfuric acid and to reprecipitate lead hydroxide or lead carbonate from lead sulfate. Lead hydroxide is more readily smelted, but a wastewater containing from 2 to 65 mg/l antimony is generated. This wastewater requires treatment to remove antimony to less than 1 to 1.5 mg/l to comply with federal wastewater regulations.

Lime or sodium hydroxide addition in the existing wastewater treatment plants has proven ineffective in antimony removal from these solutions. Addition of sodium

0097–6156/95/0607–0064$12.00/0

hydroxide to a 1,000 mg/l antimony solution (from antimony (III) chloride or antimony (V) chloride) does not produce a precipitate.

Transition metals are strongly adsorbed on metal oxides. The effects of solution pH, the concentration of transition metal, the absorbing metal oxide and the presence of other ions on the adsorption have been studied *(1,2)*. Adsorption of metals which are present in aqueous solutions as anions has had less attention in the published literature. Antimony, selenium and arsenic are metals of environmental interest which are incorporated in anionic species in aqueous solutions. Several papers have been published on adsorption and precipitation of selenium and arsenic. Little has been published on the behavior of antimony.

At pH 4.0 selenium is strongly adsorbed on amorphous ferric iron hydroxide, which is known as ferrihydrite. Adsorption is dependent on oxidation state. Selenate is completely desorbed at pH 9, and selenite is completely desorbed at pH 12 *(3,4)*. Sulfate ion interferes with selenium (VI) adsorption *(4)*. Adsorption of transition metals enhances selenium adsorption *(5)*.

Arsenic is known to react with iron oxides to generate an insoluble ferroarsenate *(6)*. Antimony may react in a similar fashion, since it is in the same group (VA) on the periodic chart.

Antimony (V) is known to be incorporated in an anionic species $[Sb_3O_9]^{3-}$ in water solution. Only in weak sulfuric acid solutions of <1.5 M does antimony (III) exist as a cation: SbO^+ and $Sb(OH)_2^+$. In 1 to 18 M sulfuric acid, antimony (III) is incorporated into an anionic species: $SbOSO_4^-$ and $Sb(SO_4)^{2-}$ *(7)*. In hydrochloric acid, antimony (III) and antimony (V) have been found to exist as a number of chloride and mixed chloride and hydroxide species, all of which are anionic *(8)*. Due to the high concentration of sulfate in the samples being treated in this study, it is likely that most of the antimony is in anionic species.

An investigation of antimony and arsenic contamination of a river bed produced some data of interest. Antimony was found to be almost entirely Sb(V) in the river water at neutral pH. River bottom samples were extracted at different pH values, and the amount of total antimony extracted was found to vary with the concentration of iron and manganese in the sediment. The total antimony extracted from the sediment reached a minimum at pH 4.3 *(9)*.

Adsorption of anions such as fluoride and phosphate on iron oxides has been investigated and found to occur at acid pH and is readily reversible at more alkaline pH *(10)*.

At pH 8.5, the average surface charge on ferrihydrite is zero. This is known as the pH_{pzc} *(11,13)*. At lower pH, the surface charge becomes more positive, and at higher pH, the surface charge becomes more negative. Anionic species are more likely to be adsorbed to a surface with positively charged sites, which are more prevalent in the lower pH range. Selenium and antimony are expected to be adsorbed at low pH.

An experimental program was designed to evaluate the possibility of adsorbing antimony from secondary smelter wastewater, to develop an industrial wastewater treatment process utilizing this adsorption, and to install the process in existing facilities.

Experimental

A sample of a wastewater collected from a secondary lead smelter is described in Table I. Unless otherwise specified, all tests were performed on this sample. Antimony (III) chloride was used to prepare antimony solutions in demineralized water. Sodium sulfate and ferric sulfate, both ACS grade, were added as needed. The solution pH was adjusted with sulfuric acid, trace metals grade.

TABLE I. TEST SOLUTION COMPOSITION

Component	Concentration	
	mg/l	Molar $(x10^{-6})$
Arsenic	15.7	210
Barium	<0.1	<0.7
Cadmium	0.16	1.4
Copper	0.28	4.4
Iron	0.45	8.1
Nickel	0.70	1.2
Lead	10.9	5.3
Antimony	95	780
Selenium	1.3	11
Tin	<0.3	<2
Zinc	1.0	15
Chloride	65	1830
Fluoride	0.4	20
Sulfate	89,900	936,000
Total Dissolved Solids	189,400	-
pH	2.5	-
Specific Gravity	1.147	-

Amorphous iron oxides were prepared by addition of ferric sulfate to the test solutions. The pH was adjusted with sodium hydroxide solution (10N) to precipitate ferrihydrite and initiate adsorption of metals. No attempt was made to control carbon dioxide or bicarbonate in the test solutions, other than to prepare fresh sodium hydroxide.

Sample aliquots were collected and immediately filtered with a 0.45 micron membrane filter. Samples were preserved for metals analysis with nitric acid, trace metal grade. Soluble antimony was determined using a Perkin-Elmer, Zeeman background corrected, graphite furnace atomic absorption spectrometer. This is a routine analysis in the analytical services laboratory, and extensive quality control records are maintained on this analysis.

The reaction time interval of 30 to 60 minutes was selected to fall within typical reaction tank retention times in industrial practice.

Results and Discussion

The pH_{pzc} is usually slightly higher for freshly precipitated ferrihydrite and decreases with aging. The adsorption edge for copper and zinc is at a higher pH for fresh precipitate than for precipitate that is a few hours old. This indicates that aged precipitate is a stronger adsorbent than fresh precipitate (*12*). For copper and zinc, the use of iron could be reduced if the iron were precipitated and aged before use as an adsorbent. Since the system is simpler if the iron is added as a solution, this loss of efficiency was accepted.

Effect of pH. Iron at 300 mg/l (5.4×10^{-3} M) was added to the wastewater described in Table I. An aliquot of the solution was adjusted rapidly to the specified pH using sodium hydroxide and held at that pH with thorough mixing for 60 minutes. The test procedure was repeated on a fresh aliquot at each pH (Figure 1). The adsorption maximum is centered at pH 5.0 and decreases at higher pH. Adsorption at higher pH may be due to a short residence time within the pH 4 to 5 range. Adsorption of antimony occurs at about the same pH as the precipitation of iron in this matrix.

Arsenic precipitation from this solution occurs at a slightly higher pH. The adsorption increases rapidly to nearly 100 percent and is not further affected by increases in pH. The contrast between these adsorption curves may indicate that the mechanism of antimony removal is not the same as that of arsenic.

Another test with sequential pH adjustment and collection of sample at each pH was performed (Figure 2). This curve provides less indication of the narrow range of pH at which the most effective adsorption occurs. The results of duplicate sample titrations above pH 6.5 are caused by slightly different rates of sodium hydroxide addition. Dependence on the rate of addition is understandable, since the degree of antimony removal is determined by the time the mixture remains in a narrow pH window.

Rate of Reaction. The rate of the reaction is critical to the design of an industrial process. 300 mg/l of iron (5.4×10^{-3} M) was added to the test solution, and the pH adjusted to 4.5 with sodium hydroxide. The reaction was nearly complete within 15 minutes, when the first aliquot was collected (Figure 3). The percent adsorbed decreased slightly after a few hours. The affinity of ferrihydrite for cadmium increased with aging (2), but decreased with aging for antimony. This effect may by due to the slow shift in pH_{pzc} of the ferrihydrite which occurs in the first few hours after precipitation (*11*).

Effect of Temperature and Concentration. Two samples were treated by addition of 100 mg/l (1.8×10^{-3} M) Fe^{+3}, sodium hydroxide addition to pH 4.5, and mixing for 30 minutes at a constant temperature. One sample was at 22°C, and the other, at

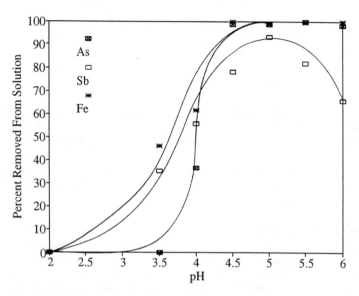

Figure 1. Effect of pH on removal of arsenic and antimony from aqueous solution by adsorption on ferrihydride.

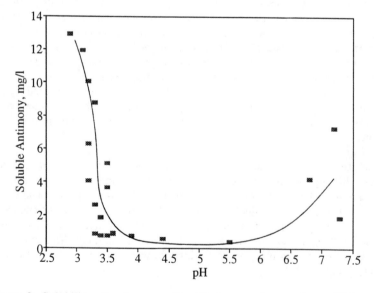

Figure 2. Solubility of antimony with pH changes due to slow addition of sodium hydroxide solution.

62°C. Soluble antimony in a sample collected from the warm sample was 5.8 mg/l (4.8 x 10^{-5} M), compared with 4.1 mg/l (3.8 x 10^{-5} M) in the room temperature sample. Antimony in the warm sample was found to decrease to 4.6 mg/l (3.8 x 10^{-5} M) after the mixture was allowed to cool overnight. An increase in temperature was found to increase adsorption of cadmium on ferrihydrite (2), but is observed to cause a decrease in antimony adsorption. This decrease is reversible upon cooling.

The influence of an increasing amount of adsorbate on the percent antimony adsorbed was investigated by adding increasing amounts of iron to aliquots of sample, adjusting the pH to 4.5 and holding that pH for 30 minutes before collecting a sample for soluble antimony analysis (Figure 4). Little improvement was found above a molar ratio of 10 moles of iron to one mole of antimony.

Interference by Sulfate. All of the testing done to this point was on an industrial wastewater sample containing about 0.95 M sulfate after ferric sulfate addition. Antimony removal from this solution was successful. To determine the effect of sulfate on antimony removal from solution, solutions of antimony, sodium sulfate, and sulfuric acid were prepared in deionized water. Aliquots of this solution were treated by sodium hydroxide addition to pH 4.5 and maintenance of this pH for 30 minutes.

At a given iron to antimony ratio, an increase in sulfate concentration resulted in a decrease in antimony removal from solution (Figure 5). Even at a molar ratio of 2.1 iron to antimony and a sodium sulfate concentration of 1.20 M, the antimony was 87 percent adsorbed. At a molar ratio of 22 iron to antimony, no change in the percent antimony adsorbed was observed with an increase in sulfate concentration to 1.2 molar. Sulfate is only a modest interference for antimony adsorption, which can be overcome by increasing the molar ratio of antimony to iron. The absence of a stronger interference by sulfate may indicate that antimony is not adsorbed on ferrihydrite on the same active surface sites as selenate. Sulfate competes for the same sites, and is adsorbed by the same mechanism, as selenate (3,5,13).

To further investigate the mechanism of antimony adsorption from solution, a sample was prepared containing 0.37 M sulfate at pH 2.0 and a molar ratio of 2.18 moles of iron per mole of antimony (100 mg/l each of iron and antimony). Sodium hydroxide was added to adjust the pH to 4.5, where it was held for 30 minutes. The pH was then increased to 9.0, mixed for 30 minutes, and an aliquot collected for analysis.

The pH of the sample was adjusted back to pH 2.0 with sulfuric acid, and antimony was added to double the initial concentration (an additional 100 mg/l). Deionized water was added as necessary to maintain a constant sulfate concentration. This cycle was repeated five times. At the end of the fifth cycle, the iron to antimony molar ratio was 0.43 or 2.3 moles of antimony per mole of iron. The antimony was over 90 percent adsorbed through the entire procedure.

Adsorption was initially postulated to be the mechanism of removal of antimony from solution. The term adsorption must be used cautiously to describe a system where 2.3 moles of antimony are removed from solution by 1 mole of iron. If adsorption is understood to be a surface effect in the absence of a precipitate, then the term has some validity.

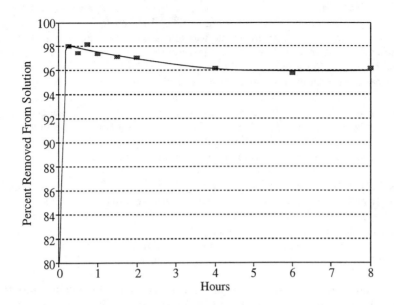

Figure 3. The rate of antimony adsorption.

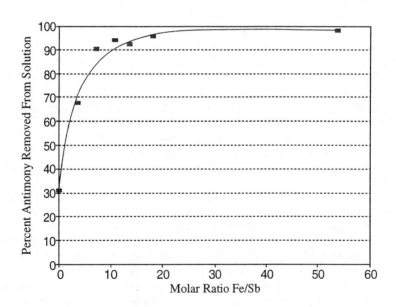

Figure 4. Percent antimony removed from solution at various iron to antimony ratios.

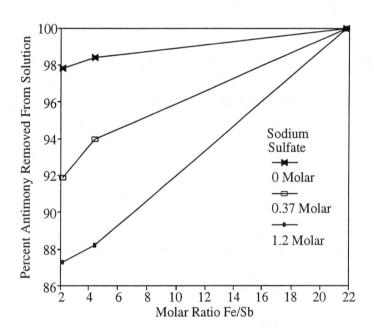

Figure 5. Sulfate interference on antimony adsorption at various sulfate concentrations and iron to antimony ratios.

The possibility of an insoluble iron compound being formed, analogous to that formed with arsenic, was considered. Neither the surface adsorption nor precipitation mechanisms seem adequate to explain the observations made above. Surface precipitation, initiated by the iron surface but not dependent on that surface once started, seems to be an adequate explanation of the observations.

To evaluate antimony desorption at alkaline pH, samples of a slurry of solids produced from treatment of wastewater by adsorption on ferrihydrite were collected. The slurry total metals and soluble metals after 24 hours of extraction of the slurry using sodium hydroxide, sodium carbonate and sulfuric acid to adjust the pH are in Table II. Antimony, once adsorbed on iron, appears to be stable and is not readily desorbed in alkaline solution. No difference was found in the desorption of antimony and arsenic.

TABLE II. SOLUBILITY OF METALS FROM FERRIHYDRITE SOLIDS IN PERCENT

	Antimony	Arsenic	Iron
Slurry (mg/l)	286	327	35900
NaOH - pH 12	0.07	<0.01	-
Na_2CO_3 - pH 11.5	0.3	0.06	-
H_2SO_4 - pH 1.2	25.4	<1	65.6

In sulfuric acid solution at pH 1.2, much more antimony was desorbed than arsenic. Antimony is only partially solubilized at low pH. No arsenic is measurable in the acidic extract. The coprecipitate between arsenic and iron is either much more difficult to dissolve or reprecipitation occurs upon dissolution.

Conclusions

A process has been evaluated which is capable of removing antimony from solution using a reasonable concentration of freshly precipitated ferric hydroxide (ferrihydrite) as an adsorbent. The pH is critical for this reaction, and sulfate interferes with the reaction to some degree. The necessary treatment reagent is available at a reasonable price, as either ferric chloride solution or ferric sulfate.

In practice, it has been necessary to extend the time required for this reaction to allow for less precision in the pH control and less agitation in the vessel used.

This process has been successfully installed at several industrial wastewater treatment plants located throughout the country. Scale up problems were mostly related to requirements for additional mixing and precise pH control to establish the pH window for antimony removal. More extensive operator training was required to enable the operators to detect and compensate for changes in the wastewater stream. Careful training and communication with the operators was found to be the most critical step in maintaining daily facility operation within compliance with regulatory requirements.

Iron addition at approximately 300 mg/l was found to be effective and sufficient to handle the routine changes in the composition of the influent stream.

References

1. Benjamin, M.M.; Leckie, J.O. *J. Colloid Interface Sci.* **1981**, 79, 208-221.
2. Johnson, B.B.; *Environ. Sci. Technol.* **1990**, 24, 112-118.
3. Hayes, K.F.; Roe, A.L.; Brown, G.E., Jr.; Hodgson, K.O.; Leckie, J.O.; Parks, G.A. *Science* **1987**, 238, 783-786.
4. Merrill, D.T.; Manzione, M.A.; Peterson, J.J.; Parker, D.S.; Chow, W.; Hobbs, A.O. *J. Water Pollut. Control Fed.* **1986**, 58, 18-26.
5. Benjamin, M.M. *Environ. Sci. Technol.* **1983**, 17, 686-692.
6. Aggett, J; Lybley, S. *Environ. Sci. Technol.* **1986**, 20, 183-186.
7. Cotton, F.A.; Wilkinson, G. *Advanced Inorganic Chemistry*, 5th ed. John Wiley: New York, **1988**.
8. *Stability Constants of Metal-Ion Complexes. Part A. Inorganic Ligands.* Hogfeldt, E., Ed. Pergamon Press: Oxford, **1982**.
9. Mok, W.; Wai, C.M. *Environ. Sci. Technol.* **1990**, 24, 102-108.
10. Machesky, M.L.; Bischoff, B.L.; Anderson, M.A. *Environ. Sci. Technol.* **1989**, 23, 580-587.
11. Stumm, W. *Chemistry of the Solid-Water Interface.* John Wiley: New York, **1992**.
12. Anderson, P.R.; Benjamin, M.M. *Environ. Sci. Technol.* **1985**, 11, 1048-1053.
13. Zhang, P.; Sparks, D.L. *Environ. Sci. Technol.* **1990**, 24, 1848-1856.

RECEIVED June 27, 1995

Chapter 7

The Use of Secondary Lead Smelters for the Reclamation of Lead from Superfund Sites

Stephen W. Paff and Brian Bosilovich

Center for Hazardous Materials Research, 320 William Pitt Way, Pittsburgh, PA 15238

The Center for Hazardous Materials Research (CHMR), in conjunction with a major secondary lead smelter, has demonstrated that secondary lead smelters may be used economically to reclaim lead from a wide range of lead-containing materials frequently found at Superfund sites. Such materials include battery case materials, lead dross, and other debris containing between 3 and 70 percent lead. During the demonstration, lead from materials from three Superfund sites and two additional sites were successfully reclaimed at the smelter. The use of secondary lead smelters to reclaim lead is anticipated to provide a viable alternative to conventional remediation techniques, which primarily work to prepare the lead for disposal.

The Center for Hazardous Materials Research (CHMR) and Exide/General Battery Corporation (Exide) performed a joint research project to determine the feasibility and economics of using secondary lead smelters for the recovery of lead from lead-containing materials.

The purpose of the project was to determine if lead from various waste materials can be economically reclaimed with minimal modifications to the process, and without any complications developing in the furnaces. Initially, the primary focus of the study was on the ability of the smelter to accept battery case materials, but the focus broadened to include other materials. CHMR and Exide processed materials from two Superfund sites containing primarily battery cases, and one battery breaker/smelter site with a variety of lead-containing materials. Two additional sets of materials, one from the demolition of a house containing lead-based paint, and the

0097–6156/95/0607–0074$12.00/0

other consisting of blasting abrasive material from work on a bridge coated with lead paint, were also processed in the smelter.

The most common Superfund materials which could be used by secondary lead smelters include materials from battery breaker and secondary lead smelter sites. The components of lead-acid batteries include: the battery case, lead electrodes (typically screens), spacers that separate the electrodes and prevent shorting, sulfuric acid, and lead battery paste ($PbSO_4$). In the past, the lead in lead-acid batteries was commonly removed by cracking or breaking the battery shell, draining the sulfuric acid into surface impoundments or lagoons, and pulling out as much of the metallic and paste lead as possible. Battery cases were often improperly disposed in drainage ditches and pits or buried with soil. Battery cases historically were made from ebonite rubber, which is a hard, black rubber containing coke and coal dusts. In the late 1970's, battery manufacturers switched to polypropylene cases, which are readily recycled.

Among the materials typically found at the 20 battery breaker sites currently on the NPL include (*1-4*): broken or whole battery cases, lead scrap, battery paste, sulfuric acid, lead debris and, (if the battery parts were burned) partially incinerated battery parts. In general, these materials are not typically fed to the secondary lead smelting furnaces, although a European smelter regularly and successfully did so during the 1970s (*5*). A major objective of the current study was to determine if such materials, after they had been sitting at Superfund sites for indeterminant periods of time, could be fed to the secondary lead smelting furnaces.

Description of Technology

The basic technology is the use of secondary lead smelting to reclaim lead from Superfund or other waste materials. The overall process involves acquiring the material, transporting it to the smelter, blending it with typical smelter feed, and then smelting it to reclaim the lead. A schematic of the process is shown in Figure 1.

Material Acquisition, Pre-processing and Transportation. The first step in reclaiming lead from Superfund wastes is acquiring and transporting the material to one of the smelters. Generally, this involves excavation or collection, pre-processing, and transport.

The lead-containing waste material is excavated from lead-acid battery Superfund sites or collected from other sources, such as bridge blasting or demolition operations.

The material requires some type of processing prior to entering the furnace. Pre-processing may include screening to remove large debris or soil, and analyses to determine composition and lead content. Pre-processing can be performed at the site or at the smelter, depending on which is most cost efficient.

Because secondary lead smelters are fixed facilities, it is necessary to transport the lead-containing waste material to the smelter. The material is usually considered hazardous due to lead content, so the materials must be handled by licensed carriers that are permitted to carry hazardous waste.

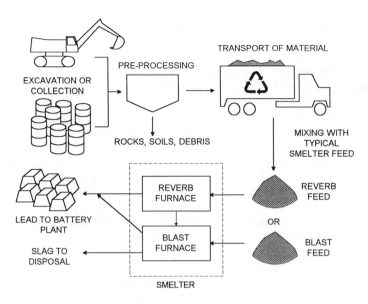

Figure 1. Schematic of Reclamation Process

Mixing with Typical Feed. Once the material arrives at the secondary lead smelter, it is blended with typical feed prior to processing through the furnaces. Anticipated feed ratios range from 10% to 50% by weight, and are determined based on evaluating furnace performance when testing some of the feed. At Exide's smelter, the lead-containing materials can either be processed through the reverberatory furnaces, or the blast furnaces. In general, smaller sized feed is processed through the reverberatory furnaces and larger material is processed through the blast furnaces.

Smelting Process. Figure 2 shows a process schematic for Exide's secondary lead smelter, which is typical of the secondary lead smelting industry. Exide is primarily engaged in recycling lead from spent lead-acid batteries and other lead containing materials. Exide reclaims thousands of tons of lead annually using both reverberatory and blast furnaces. All of the evaluations described as part of this research effort were conducted at Exide's Reading, PA, smelter.

Spent batteries received at the Reading smelter are crushed in a Saturn shredder. The crushing process releases the sulfuric acid, which is collected and piped to an on-site wastewater treatment facility. The crushing also reduces the polypropylene battery cases to chips. The lead and polypropylene are then processed through a hammermill into a sink/float process, where the lead is physically separated and conveyed to feed piles within a containment building. The polypropylene is collected and processed through the facility's plastic recycling operation, where they are thoroughly cleaned and used in the production of new battery cases. In general, battery cases are not fed to the furnaces, but are recycled, and only the lead scrap is fed to the furnaces.

The two reverberatory furnaces at the Reading smelter are charged with material from the sink/float system as well as other lead containing material. These furnaces are fueled with natural gas and oxygen. Material is charged to the reverberatory furnaces by a conveyor system that feeds into a ram which pushes material into the furnace. These furnaces are tapped for slag, which typically contains 60 to 70 percent lead, and a soft (pure) lead product.

The two blast furnaces are charged with the slag generated from the reverberatory furnaces as well as other lead-containing materials. These furnaces are fueled by coke, iron, air and oxygen. The blast furnaces are charged with 3 to 6 tons of material per hour. Iron and limestone are added as fluxing agents to enhance furnace production. The blast furnaces are tapped continuously to remove lead and intermittently to remove the slag. The blast slag, which contains primarily silica and iron oxides, is disposed at an off-site landfill.

Lead produced in the blast and reverberatory furnaces is transferred to the refining process where additional metals are added to make specific lead alloys. The lead is then sent to the casting operations where it is molded into "pigs." These are transported to Exide's battery manufacturing facilities for use in the manufacture of new lead-acid batteries.

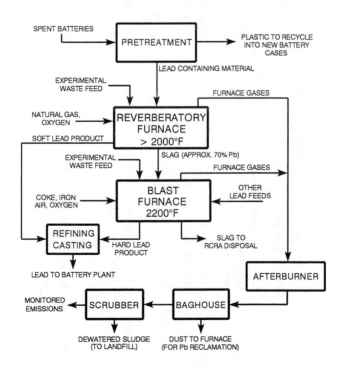

Figure 2. Schematic of Secondary Lead Process

Table I. Summary of the Evaluations

Site	Site Description	Site Size	Test Duration	Type of Mar'l	Amt. Processed	Avg. Lead Conc.	Feed wt. Ratio
Tonolli	Integrated battery breaker	30 acres	5 days	Rubber & plastic battery cases	84 tons	3.5%	20%
Hebelka	Auto junk & salvage yard	20 acres	1 day	Rubber & plastic battery cases	12 tons	14.7%	17%
Laurel House	Residence	N/A	1 day	Wood demolition debris	4 tons	1%	5%
NL Industries	Integrated battery breaker & smelter	11 acres	3 months	lead slag, debris, dross, ingots, cases, baghouse dusts	1570 tons	30 to 60%	20 to 50%
PennDOT	Bridge repainting site	N/A	1 day	Iron shot bridge abrasive	6 tons	3.2%	13%

Study Methodology, Sites and Materials

The basic methodology during the project was to acquire quantities of the materials to be tested, characterize these materials, process them through the secondary smelter, and evaluate the effects on the smelter. Materials from three Superfund sites as well as two additional sets of lead-containing materials commonly produced were processed during this project. The following sections provide a short description of each of the five sites and evaluations. The feed rate is given in percent by weight, and is calculated by dividing the weight of the test material charged to the furnace by the total feed weight including the test material. Table I presents a summary of the evaluations.

Tonolli Superfund Site. The Tonolli site was a 30 acre battery breaking and smelting facility located in Nesquehoning, PA. The material consisted of piles of battery case pieces at the site, so the material could be loaded directly onto the trucks. The material was fed at a ratio of 10% by weight of test feed and 90% by weight regular feed through one of the smelter's reverberatory and blast furnace combinations. Due to the large size of the material, it was not successfully processed through the reverberatory furnace.

Hebelka Superfund Site. The Hebelka site was a 20 acre automobile junk and salvage yard located in Weisenburg Township, PA. The site contained battery case debris mixed with soil that had an average lead concentration of 14.7%.

The first step was to reduce the material in size from 1 through 14" pieces to less than ¼" pieces using a hammermill. The test material was mixed with regular feed to produce a feed ratio of 17% test material, by weight, as fed to the reverberatory furnace.

Laurel House. Laurel House was a women's shelter located in Montgomery County, PA. Most of the interior and exterior of the house was coated with lead based paint, which was removed and replaced. The material shipped to Exide consisted mainly of wooden demolition debris from the rear porch contaminated with the lead paint. The material had a lead composition between 0.5 and 1%. The material was reduced in size with a pallet shredder as it arrived at the smelter.

The demolition debris was processed through both reverberatory furnaces at a feed ratio of 10% test material, by weight. Although this is a low ratio, the test material comprised 50% of the volume of feed to the furnaces. The feed ratio was later reduced to 5% to reduce the mechanical malfunctions in the furnaces. After the feed ratio was adjusted, the furnaces seemed to perform normal.

NL Industries Superfund Site. The National Lead site was an 11 acre integrated battery breaking, smelting, and lead refining facility with its own on-site landfill. There were a wide variety of materials at the site including lead slag, dross, debris, ingots, hard heads; battery case debris, baghouse bags, and contaminated pallets and iron cans. This evaluation was conducted in two parts: a preliminary investigation and a full-scale investigation.

During the preliminary investigation, approximately 370 tons of all types of the above materials were processed. Analyses revealed an average lead concentration of 57%. The larger pieces of debris were sorted out and processed through one of the blast furnaces, while the bulk of the material was fed into one of the reverberatory furnaces at feed ratios of up to 100%. This amount of material caused many malfunctions in the feed system to the furnaces, so the ratio was lowered to 50% test material, by weight. Exide personnel did not notice any further complications in the reverberatory furnaces after this change.

During the three months of the full scale operation, approximately 1,200 tons of material were transported to Exide's Reading smelter. For the first two months, the test material was processed in the one of the smelter's reverberatory furnace. The feed ratio was 20 to 30% due to the amount of calcium in the test material. Excess calcium causes buildups and other problems in the furnaces. The lead concentration during this part of the investigation averaged approximately 50%.

The test material for the last month of the investigation was mainly larger pieces of slag and debris with an average lead concentration of 30%. This material was charged directly to one of the blast furnaces, at a feed ratio of approximately 30%.

Pennsylvania Department of Transportation. The Pennsylvania Department of Transportation (PennDOT) used an iron-shot abrasive blasting material to remove old lead-based paint from a bridge in Belle Vernon, PA. Exide processed sixteen 55 gallon steel drums at the Reading smelter. The material contained an average of 3.2% lead and approximately 50% iron.

Exide personnel determined that the material contained too much moisture to be incorporated into the reverberatory furnace feed system. The test material, including the iron drum, was then fed to one of the smelter's blast furnaces at the rate of about one per charge, with the furnace being charged approximately twice every hour. The feed ratio was approximately 13% test feed, by weight.

Furnace Assessment Methodology

CHMR personnel were on-site at Exide during most of the evaluations to monitor furnace performance. This monitoring included the following: recording furnace parameters, taking samples, observing mixing and feeding procedures, and acquiring daily production sheets. Exide has two sets of reverberatory and blast furnaces. Changes in the parameters were determined by monitoring the furnace charged with experimental feed (the "test" furnace), as well as the parallel furnace charged with regular feed (the "control" furnace).

Furnace Parameters. Certain furnace parameters, such as air feed, oxygen uptake, furnace back pressure, and sulfur dioxide emissions, were monitored directly by CHMR personnel to assess furnace performance. The data obtained was helpful in determining the effects of the material, and additional processing costs for that material.

Sample Collection. Samples were collected over the course of many of the evaluations. These included test material feed samples, reverberatory and blast furnace slag samples, sludge samples, and lead samples. CHMR retained custody of feed samples, slag samples, and sludge samples. These were analyzed by independent laboratories, and the lead samples were analyzed for trace metal and alloy content on-site at Exide's laboratory. Sampling the test furnace and control furnace allowed comparisons to be made between typical furnace production and the production when test material was being fed.

Material Processing Oversight. CHMR personnel were on-site at the smelter to observe the procedures used to pre-process, mix, and feed the test material, when applicable. The observations were helpful when determining the cause of any furnace problems or malfunctions that may occur during the evaluations, and to determine the costs associated with processing the test material.

Review of Daily Production Data. The daily production sheets prepared by Exide are a summary of the types of material processed, furnace problems, lead and slag production, and many other events that occur during the course of the day. Comparisons between test furnace and control furnace production can be made because these sheets provide the amount of slag and lead produced on a daily basis. The sheets detail any furnace malfunctions that may have occurred, and they indicate the number of hours each furnace was "down" each day. A furnace is considered down when the production through it is stopped so that the furnace can be repaired.

Results and Discussion

The investigation demonstrated that it was feasible to reclaim lead from all of the test materials processed through the secondary lead smelter. This section details the analysis for one test using material from the NL Industries site in Pedricktown, NJ, which typifies both the types of material and the kinds of effects noted during the evaluations.

NL Industries Site Evaluation. Lead contaminated waste material was processed from the NL Industries site in two phases. During the first, preliminary phase, approximately 370 tons of material were transported from the site to Exide's smelter in Reading, PA. The purpose of the preliminary phase was to determine the best method to mix, feed, and process the material, at what rate, and the affects on the furnace. During the second phase of the test, an additional 1,200 tons of material were removed from the site and processed at the Reading smelter.

During the preliminary investigation, the material was processed through one reverberatory furnace at various feed rates (30, 50 and 100% of furnace feed). The material was well-suited for processing through the reverberatory furnace, in that it had moderately high levels of lead, however, it was denser than normal furnace feed. When fed to the furnace at 100% of the feed rate, it caused the conveyor belt entering the furnace to tear and the ram to jam, temporarily shutting down the furnace.

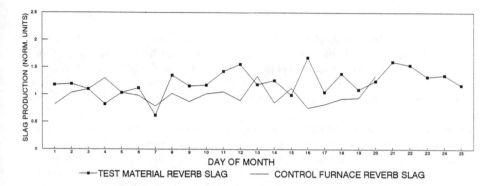

Figure 3. Comparison of Slag Production in the Test and Control Furnaces During August 1992

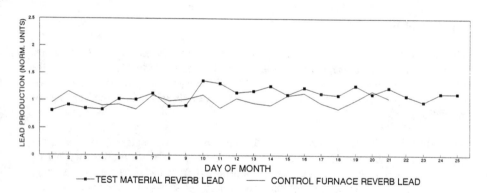

Figure 4. Comparison of Lead Production in the Test and Control Furnaces During August 1992

Therefore, the feed ratio was reduced to 50% and below, where the furnaces experienced no malfunctions that could be attributed to the test material.

The furnace parameters monitored during the preliminary test included oxygen and natural gas feeds, production, and sulfur dioxide emissions. The test material required an increase of approximately 120 cubic feet per minute (cfm) oxygen and 30 cfm natural gas in the test reverberatory furnace over the control furnace. This was anticipated because the material contained more slag and silica than normal feed, and had a lower Btu value. Therefore, the material required slightly more energy to process.

Sulfur dioxide concentrations in the stack gases did not differ significantly from those of the control furnace, and were within Exide' permitted compliance limits. This was expected, because the sulfur dioxide concentrations are controlled by a scrubber. Also, the material contained less than 3% sulfur, which is the concentration of sulfur in the regular furnace feed. In addition, no increase was found in the amount of calcium sulfate sludge generated in the scrubber, further indicating that no excessive sulfur was fed to the system.

During the longer term evaluation, in which 1,200 tons were processed in the smelter, approximately 800 tons were processed through the reverberatory furnaces. The materials were fed to the furnace at approximately a 30% feed ratio over a two month period. CHMR analyzed the lead and slag production data over this period and concluded that the use of the material did not decrease the production of either lead or slag from the furnace.

Figures 3 and 4 show a comparison between the reverberatory slag and lead production for the test and control furnaces during August 1992. There are wide variations in daily production, but no consistent pattern of one furnace exceeding the other. The pattern during September is similar.

The blast furnaces were fed the material from the NL Industries site during October 1992. When furnace parameters were evaluated over that month, the results showed no differences in most furnace parameters, including coke and air requirements, lead production and waste production.

Overall Results. In general, the evaluation demonstrated that the tested materials may be processed in secondary lead smelters with relatively few effects on overall furnace production. The most significant effects were caused by processing too much material at one time, or by processing materials in a furnace without proper pre-processing. For example, the Tonolli feed was too large to be processed effectively in the reverberatory furnaces, and caused the furnace production to slow down significantly when processing was attempted. Later, a similar material, from the Hebelka site, was successfully processed in the reverberatory furnace after it had been shredded so that the particle size reduced to less than ¼ inch. Similarly, the NL Industries site material was successfully processed only at feed ratios below 50%.

Some materials, particularly battery cases, represent a potentially significant source of energy to the furnaces. The battery case materials tests had average Btu-values of over 11,000 btu/lb. The relatively high heating values of these materials is not surprising since many battery cases (particularly those which are 15 years or more old) are made of ebonite rubber, which contains significant amounts of coal and coke

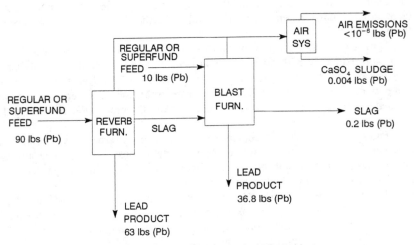

BASIS: 100 lb total lead input: 90 lbs to reverb, 10 lbs to blast

Figure 5. Materials Balance for Lead Over the Secondary Lead Smelter

dust. Coke is a feed to a blast furnace, in which it acts to reduce metal oxides to their metallic forms. Ebonite rubber cases were successfully substituted for a portion of the coke in the blast furnace. Battery cases will not substitute for all of the coke, however, since the large chunks of coke also provide structure inside the furnace, which battery cases cannot provide. The reduction in coke btu requirements from battery cases is difficult to determine, but based on discussions with the furnace operators, it was estimated at between 2,500 and 3,000 btu/lb.

Other materials represent substitutes for other furnace feeds. Iron is fed to a blast furnace to help separate lead from the slag. The PennDOT bridge blasting material represented a source of iron in the blast. The material contained so much iron (over 60%) and so little lead in comparison (approximately 3%), that it was considered an iron source which happened to be contaminated with lead, rather than a lead source to the furnace. In this capacity, it represents a potentially beneficial reuse for a material which most state and municipal authorities have found difficult to dispose. Use of the PennDOT material as an iron source did not have a measurable effect on any furnace parameters.

Materials Balance. In determining whether the process is suitable for the Superfund type materials, it is important to determine if the secondary lead smelting process actually reclaims lead from the various materials, and the extent to which the lead is released back to the environment.

Figure 5 shows a materials balance for lead throughout the process. It is based on feeding 100 pounds of lead, (90 lbs to the reverberatory furnace, and 10 lbs to the blast). The figure shows that lead may enter either the blast or reverberatory furnaces, and has outlets in the lead produced, slag, calcium sulfate sludge and monitored stack emissions. No other outlets of lead exist under normal plant operations. The figure indicates the percentages of lead entering the furnaces which end up in each outlet stream. As can be seen from the figure, the furnaces reclaim approximately 99.7% of the lead fed to them.

The reclamation efficiency for Superfund materials is primarily determined by the percentage of lead in the feed, and the percentage of the material which will end up in the slag. If the Superfund materials were not significantly different from other feed materials, then their reclamation efficiencies would be similar to those of other streams: 99.7%. However, many of the Superfund materials contain less lead, and greater concentrations of ash which end up in the slag. This tends to decrease the reclamation efficiency, as the slag typically contains 2 to 4% lead (6), which will be drawn from the feedstocks. The reclamation efficiencies for the Hebelka and NL Industry materials are estimated to be above 99%. The reclamation efficiency for the PennDOT material, in contrast, is estimated at approximately 40%, since that material contained a relatively low percentage of lead (3%), and a high percentage of material which ended up in the slag (60% iron). For the purposes of estimating reclamation efficiency, the slag is estimated to contain 3% lead (6).

Economics

The cost of using secondary lead smelters for the recovery of lead from Superfund sites is determined by the cost of excavation, transportation and processing at the smelter. Excavation costs for Superfund materials vary depending on the material type, location and accessibility. Excavation costs will increase if additional on-site processing is required before the material can be transported. Transportation costs to the fixed smelter location depend on the amount of material transported and are directly related to the distance transported.

Processing costs for the materials were found to vary significantly depending on the concentration of lead, the market price for lead, and the percentage of the feed which becomes slag. As the market price for lead or the concentration of lead in the feed material decreases, the cost of processing Superfund materials will increase, because the lead represents a salable commodity generated during the reclamation. If the material contains a greater fraction of constituents which exit the furnace in the slag fraction, then the cost increases commensurate with the disposal costs of slag. Most of the other parameters (for example, a slight increase in oxygen usage in the furnace, or the Btu-value of the feed) have little overall effect on the cost of processing.

Conclusions

Based on the experiments performed, CHMR concludes that the use of secondary lead smelters for the reclamation of materials from Superfund sites is a technically and economically viable option. The results from this study do not guarantee that the technology is applicable to other types of material found at other sites because the sites may contain varying concentrations of lead and other metals. CHMR and Exide believe that a feasibility study may be required for each material to determine which are suitable for this process, the required furnace configurations, feed rates, and economics.

Acknowledgments

This project was partially funded under a grant from the U.S. Environmental Protection Agency Office of Research and Development (contract CR-818199). The additional support from Exide/General Battery Corporation, including monetary, facility, and personnel support, is gratefully acknowledged.

Literature Cited

(1)   Royer, M.; Basu, T. *Selection of Control Technologies for Remediation of Lead Battery Recycling Sites*, EPA Risk Reduction Engineering Laboratory, Edison, NJ, 1991.

(2)   Barth, E.; Soundararajan, R. *HMCRI Superfund '90 Conference*, Hazardous Materials Control Research Institute, Greenbelt, MD, 1990, p. 665.

(3)   Royer, M.; Selvakumar, A.; Gaire, R. *J. Air and Waste Management*, Vol 42, No. 7, 1992, p. 970.

(*4*)   Hessling, J. et al. *2nd International Symposium on Metals Speciation, Separation and Recovery*, Rome Italy, 1989, p. 183.

(*5*)   Coleman, R.; Vandervort, R., U.S. EPA Office of Research and Development, EPA-600/2-80-022, 1980.

(*6*)   Queneau, P.; Cryar, D.; Mickey, D. In *Primary and Secondary Lead Processing*; Jaeck , R. Ed; The Metallurgical Society/Pergammon Press, New York, 1989.

(*7*)   Isherwood, R.; et al., *The Impact of Proposed and Existing Regulations Upon the Domestic Lead Industry*; U.S. Bureau of Mines, Denver, CO, 1988.

RECEIVED June 2, 1995

Chapter 8

Recovery of Phosphates from Elemental Phosphorus-Bearing Wastes

Ronald E. Edwards, Jack M. Sullivan, and Oscar E. Moore

Environmental Research Center, Tennessee Valley Authority,
P.O. Box 1010, Muscle Shoals, AL 35662-1010

A process for oxidizing aqueous elemental phosphorus containing residues (sludges) to produce orthophosphate containing slurries suitable for subsequent reaction with ammonia to produce nitrogen and phosphate containing plant nutrient products. It comprises reacting aqueous elemental phosphorus containing residues with certain special mixtures of concentrated nitric acid and sulfuric acid to effect the conversion of the elemental phosphorus into mostly orthophosphoric acid and very little orthophosphorus acid with relative ratios of the two acids being dependent upon the mole ratio of $H_2SO_4:HNO_3$ employed in the processing. The resulting aqueous reaction intermediate is neutralized with ammonia during processing to a fluid or solid plant nutrient product. Prior to the conversion to products, the aqueous reaction intermediate may be subjected to a solids separation step to remove insoluble salts of certain environmentally undesirable metals, such as Pb, Cd, Ba, and Cr.

Historically high purity elemental phosphorus has been used for the production of a number of industrial products including detergents, dentifrice products, food additives, phosphatizing agents, fertilizers, and a whole host of phosphorus containing chemicals (1). During war time, huge quantities of elemental phosphorus is required for the production of ammunitions and incendiary bombs (2). Practically all elemental phosphorus was produced by the electric furnace method which also produced quantities of elemental phosphorus bearing wastes. In recent years most of the furnaces were decommissioned and some of the above mentioned uses of elemental phosphorus have been replaced by less pure phosphates produced from purified wet process phosphoric acid. However, large quantities of the elemental phosphorus bearing wastes still exist at most of the operational and non-operational phosphorus furnace sites, military installations, and industrial sites that used elemental phosphorus as a raw material. As of 1992, there was an estimated total of 400 acre-feet of these phosphorus bearing wastes in the form of sludges awaiting treatment in the United States (3).

The objective of this research was to develop an environmentally sound process to safely process elemental phosphorus bearing wastes by minimizing or eliminating all processing by-products and producing a usable, and hopefully marketable, product other than elemental phosphorus. This process development study was initiated by the Tennessee Valley Authority (TVA) to develop technology for the processing of elemental phosphorus containing by-products produced by the electric furnaces operated until 1976 at Muscle Shoals, Alabama. This technology could be useful at numerous other facilities throughout the United States that once produced elemental phosphorus by the electric furnace process or utilized phosphorus as a raw material.

Background of Phosphorus Wastes

In the electric furnace process for producing elemental phosphorus, agglomerated phosphate-containing ore (apatite) is reacted with coke at temperatures of 1200 to 1500 °C. Silica is added to the reaction mixture to serve as a flux to remove calcium silicate slag. A simplified equation for the overall process is shown below (4).

$$2 \, Ca_3(PO_4)_2 + 6 \, SiO_2 + 10 \, C = P_4 + 10 \, CO + 6 \, (CaO\text{-}SiO_2) \qquad (1)$$

The gaseous phosphorus and carbon monoxide, along with impurities (generally SiF_4 and dust), are passed to an electrostatic precipitator where a major portion of the dust is removed at temperatures above the dew point of phosphorus. The gaseous phosphorus and carbon monoxide then proceed to condensing columns where the phosphorus is condensed to a liquid by water sprays maintained at 45-55 °C. The uncondensed carbon monoxide exits the sprayers and may be burned as a fuel. Ferrophosphorus is tapped from the furnace as a by-product for use in steel production. The calcium silicate slag is removed as a waste product containing no elemental phosphorus.

The water-covered liquid phosphorus runs to a sump where a small amount of impurity sludge collects at the interface between the water and the liquid phosphorus. This sludge layer consists primarily of hydrated silica, fluorosilcates, dust, and emulsified phosphorus. A portion of the entrained phosphorus is recovered from the sludge by centrifugation and a limited amount of the sludge is recycled to the furnace. However, the physical and chemical properties of the residue prevents its complete recycle and the remaining sludge must be treated as a waste or processed by some other means. In past practice, phosphorus producers often stored such by-product sludges in ponds, sumps, or tanks for future processing. Due to certain environmental considerations associated with the chemical properties or characteristics of elemental phosphorus, such practices are now discouraged and methods for the complete recovery, or conversion of the elemental phosphorus values in such sludges to benign and useful products, are needed (2,4,5).

The elemental phosphorus containing sludge is a very non-homogeneous material varying in consistency from thin slurries to a thick mud. Upon aging hard crystalline layers may also form. Sludges that are newly formed can contain elemental phosphorus concentrations as high as 50-60% by weight. However, sludges that have settled for several years (which includes most sludges stored at the old furnace locations) will have elemental phosphorus concentrations of generally 5-15% by weight. These sludges usually contain small amounts of various heavy metals including iron, aluminum, lead, arsenic, barium, cadmium, and others. A typical analysis of phosphorus sludge located at Muscle Shoals, Alabama, is provided (Table I).

Table I. Typical Analysis of Phosphorus Sludge

	Concentration	
	Mean	Range
Total P, wt %	20.2	6.1 - 37.5
Elemental P, Wt %	11.0	0.7 - 32.9
Fe, wt %	2.6	1.0 - 5.5
C, wt %	2.0	0.9 - 3.4
Al, wt %	0.2	0.1 - 0.4
Pb, ppm	310	70 - 600
As. ppm	42	20 - 70
Ba, ppm	19	0 - 40
Cd, ppm	30	0 - 60
Hg, ppm	<1	<1

Over the past 25 years the number of electric furnaces used to produce elemental phosphorus has decreased at a rapid rate. A large portion of the electric furnace phosphorus was used to produce a pure grade of phosphoric acid and munitions. With the development of improved methods of purifying wet-process phosphoric acid and the increased energy and environmental costs of operating electric furnaces, the furnaces began to close down in the 1970's and 1980's. In 1968 there were 36 furnaces in operation in the United Sates at 13 locations producing 655,000 t/y. By 1992 there were only eight operating furnaces in three locations with an estimated production of 265,000 t/y (4). Large quantities of elemental phosphorus containing sludge and precipitator dust have been stored at most locations of both operational and non-operational phosphorus furnaces. At most non-operational furnace locations and manufacturing facilities that have stored elemental phosphorus as a feed stock for various industrial processes, little or no expertise or equipment is available to properly handle, transport, or store the hazardous sludge. For this reason a process that could be used on location to convert the elemental phosphorus first to environmentally benign phosphate compounds and then to a usable, marketable product would be beneficial.

TVA operated an electric furnace process producing elemental phosphorus at Muscle Shoals, Alabama, from 1934 to 1976. Although TVA made major efforts to recycle the sludge produced by its furnaces, there was still a large amount of phosphorus sludge at the time operations were terminated. In the last few years prior to the shutdown of the furnace operation, TVA stored approximately 130,000 liters of the phosphorus sludge in rail cars, and tanks for future process studies. In 1990 a multidisciplinary team was established to evaluate existing processing options to utilize or convert the sludge to a useful product, as well as develop new technology alternatives.

Remediation Alternatives

The remediation alternatives for phosphorus sludge that were evaluated during this preliminary study included incineration, distillation, air/oxygen oxidation, long term containment, recycling into a new or existing furnace, and nitric acid oxidation. The major disadvantages associated with most of the above mentioned remediation methods included landfilling process by-products, incomplete conversion of the elemental phosphorus, handling and shipping hazards associated with an elemental phosphorus containing product, high capital costs, and insufficient testing of technology.

After a lengthy evaluation of all available options, it was decided that technology based on nitric acid oxidation of the sludge followed by ammonia neutralization to produce a nitrogen and phosphate containing plant nutrient was the most advantageous approach. It avoids most of the disadvantages mentioned above and this type process would offer the advantages of producing a safer phosphate product and would produce no by-product requiring disposal in a landfill. Also, this process alternative fit the technical expertise at TVA's Environmental Research Center in Muscle Shoals, Alabama, and offered the type of technology most likely to be accepted by environmental regulators and the public. TVA was prepared to initiate a study utilizing its 130,000 liters of stored phosphorus sludge to develop and demonstrate this technology.

The proposed process was based upon the assumption that the elemental phosphorus in the aqueous sludge could be oxidized with nitric acid to produce a phosphoric acid solution which in turn might be ammoniated to produce an ammonium phosphate containing mixture suitable for granulation to a fertilizer product. A simple flow diagram of the overall process proposal is provided (Figure 1). The assumed chemical reactions are as shown:

$$20\ HNO_3\ +\ 3\ P_4\ +\ 8\ H_2O\ =\ 12\ H_3PO_4\ +\ 20\ NO \tag{2}$$

$$12\ NH_3\ +\ 12\ H_3PO_4\ =\ 12\ (NH_4)H_2PO_4 \tag{3}$$

These equations assume that the phosphorus is converted to orthophosphate and that the nitric acid is reduced to nitric oxide, NO. However, in calculating the stoichiometry of this system it must be noted that the nitric acid can be reduced to one or a combination of four gaseous reduction products: nitrogen dioxide, NO_2; nitric oxide, NO; nitrous oxide, N_2O; and nitrogen, N_2. Also, the oxidation may not proceed all the way to the orthophosphate oxidation state and may stop at the orthophosphite

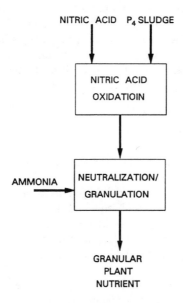

Figure 1. Flow diagram of TVA remediation process.

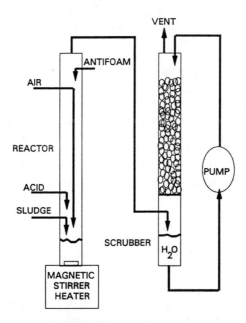

Figure 2. Lab-scale batch oxidation reactor.

(H$_3$PO$_3$) or the hypophosphite (H$_3$PO$_2$) oxidation stages (6,7). Without any quantitative information concerning the extent to which each of the mentioned reactions might occur, it was decided to base all initial calculations upon the assumption that the oxidation reaction proceeds according to eq. 2.

Lab-Scale Study

In early 1991 a laboratory investigation was initiated to develop the nitric acid oxidation technology. A 55-gallon drum of representative phosphorus sludge was collected from one of the storage tanks and was found to contain 7.5-7.9% elemental phosphorus. (The nonhomogeneous nature of the sludge was responsible for the range of analysis.) The lab-scale test assembly shown in Figure 2 consisted of a reactor and a scrubber. Both vessels were cylindrical Pyrex glass 90 mm ID x 122 cm total height. Nitric acid, sludge, antifoam agent, and air were all added through glass fittings at various locations on the side of the vessel. The reactor contents were heated and agitated by means of a magnetic stirrer-heater. Reactor temperatures were measured by a type K thermocouple. The reactor was designed to process approximately 1 liter of sludge per batch. Gases evolving from the reactor were carried by 8 mm ID glass tubing to the scrubber vessel. The scrubber contained 80 cm of 1-inch stainless steel Pall packing and water which was recirculated over the packing by a peristaltic pump. Preliminary beaker tests had shown the reaction to produce large amounts of foam. For this reason the reactor was designed with a large freeboard and allowances for an antifoam agent addition.

In the initial tests, 125-150% of the stoichiometric nitric acid was fed to the reactor using both 57% and 70% HNO$_3$. The large excess of nitric acid was to ensure the complete oxidation of the elemental phosphorus and to compensate for other impurities in the sludge that might compete for the oxidizer. The earlier tests of this series showed that the foaming problem was more easily controlled when the batch reactor was precharged with the nitric acid and the sludge was metered into the acid over a 30-45 minute period. Reaction temperatures ranged from 102-110 °C and resulted in continuous boiling of the reaction solution, with the evolution of large quantities of NO$_x$ gases. After all the sludge was fed, the reactor temperature was held at the reaction temperature for a retention time of 2 hours. Results of this first group of tests are shown in Table II. The product slurry contained only 5-10 ppm of elemental phosphorus; however, the orthophosphate was only 47-57% of the total phosphorus and the remaining phosphorus was orthophosphites. No hypophosphite was identified. These levels of orthophosphites presented a problem in converting the oxidized slurry to a plant nutrient. Unlike orthophosphates, orthophosphites are toxic to plants at high concentrations and at the above mentioned levels, orthophosphites would nullify the effect of the plant nutrient properties of the orthophosphates.

In the next series of tests varying amounts of sulfuric acid were added to the 125-150% stoichiometric equivalent of 56% nitric acid in the reactor. These acid mixtures were used in hopes of increasing the oxidation level of the phosphorus and reducing the orthophosphites to levels that would not adversely effect the plant nutrient value of the product. Weight ratios of H$_2$SO$_4$ (93%) to HNO$_3$ (56%) were tested at 1:3, 2:3, 1:1, and 3:2. Results of these tests are shown in Table III. The mixed acids reduced levels of elemental phosphorus (0.8-3.2 ppm) at all acid ratios and resulted in a substantial increase in the production of orthophosphates relative to orthophosphites when the acid weight ratio was above 1:1. Figure 3 shows a graphical representation of the effect that the ratio of sulfuric to nitric acids had on the conversion of elemental phosphorus to orthophosphates (6,7).

Table II. Results of Lab-Scale Tests

	57 % Nitric Acid	70 % Nitric Acid
Sludge		
P_4, wt %	7.9	7.5
Total P, wt %	9.8	9.0
Oxidation Product		
P_4, ppm	5.9	9.9
Total P, wt %	4.9	6.3
Orthophosphates, wt % of total P	57	47
HNO_3, wt %	21	12

Table III. Lab Results Using Sulfuric/Nitric Acid Mixtures

	Weight Ratio of H_2SO_4 : HNO_3 [a]			
	1:3	2:3	1:1	3:2
Oxidation Product				
P_4, ppm	1.3	3.2	0.8	1.8
Total P, wt %	5.2	4.6	4.5	3.7
Orthophosphate, wt % of total P	50	53	93	98

[a]H_2SO_4 - 96 %; HNO_3 - 56%

The results of the lab-scale study proved that the elemental phosphorus in the sludge could be effectively oxidized using a mixture of sulfuric and nitric acids with a weight ratio of 1:1 or higher and a 25-50% excess nitric acid. The reactor design requires a 300% freeboard to adequately control the foaming even with the use of a foam retardant. The reaction can be controlled at the boiling point of the solution (105-115 °C), thereby allowing evaporative cooling to remove excess heat from the reaction. Based on these results it was decided to build a bench-scale plant (20:1 scale-up) to further optimize the oxidation process, evaluate the effects of scale up, and produce enough product to evaluate the production of a nitrogen and phosphate plant nutrient product.

Bench-Scale Development

In mid-1992 a bench-scale facility designed to process approximately 20 liters of phosphorus sludge per batch was constructed and successfully operated. The facility consisted primarily of a phosphorus sludge feed tank, a feed pump, and a reactor. The sludge feed tank was a 316 stainless steel cylindrical cone bottom tank, 66 cm in diameter by 100 cm high, containing a turbine-type agitator. The feed pump was a peristaltic metering pump. The reactor was constructed of 316 stainless steel with dimensions of 25 cm ID by 182 cm high. The lower 30 cm of the reactor was jacketed for heating and cooling. The reactor was baffled and fitted with a turbine type-agitator and had J-type thermocouples located throughout the reaction zone. No scrubbing system was added to this facility due to the expected short duration of the tests. However, an exhaust system was designed to allow for air sampling that would be necessary to design an adequate scrubber system for any additional testing or scale-up. A diagram of the bench-scale oxidation plant is shown in Figure 4. The process procedures were designed from the optimum results of the lab study with minor modifications to adjust for scale up and potential industrial applications.

In the lab-scale tests the feed rates were based on the assumption that all phosphorus was oxidized to orthophosphates and all the nitric acid was reduced to nitric oxide (eq 2). For the bench-scale tests the assumption that all phosphorus would be completely oxidized to orthophosphates was accepted. However, it was now believed that a mixture of the four nitric acid reduction products mentioned previously is formed. Therefore, an equally weighted average of the four nitric acid reduction stoichiometries was used to determine feed rates. Based on this assumption, the mole ratio of HNO_3: P_4 was calculated to be approximately 8.9. When an excess of 25-50% nitric acid is incorporated into the formulation, the HNO_3:P_4 mole ratio is set for a range of 11.1 to 13.4. The sulfuric acid feed rate was based on the results correlated in Figure 3 and set at a range of 1.1 to 1.3 mole ratio of sulfuric acid to nitric acid.

The results of the bench-scale tests confirmed the results obtained in the earlier lab-scale tests, and process procedure modifications were developed that would assist in further scale up of the process. Oxidation of the phosphorus sludge was excellent with the product containing 0.3-3.0 ppm elemental phosphorus and 97-99% of the total phosphorus was in the form of orthophosphate. The bench-scale oxidation results are given in Table IV. The scale up of the reactor created a decrease in the ratio of surface area to volume and caused the reaction temperature to increase to 120-124 °C in some of the tests. Reaction temperatures above 120 °C produced excessive volatilization and/or thermal reduction of the nitric acid and reduced the oxidizing capacity of the reactor. To control the reactor temperature, many procedures were attempted or theorized that would correct this problem. These included reducing the sludge feed rate; reducing the concentration of the elemental phosphorus in the sludge by dilution; adding a portion of the nitric acid feed (10-15%) after the completion of the sludge feed; installing a cooling jacket or coils to control the reaction temperature; and adding a condenser to the reactor vent to return the volatilized nitric to the reactor. The two procedures found to be most practical for this test series were slowing the feed rate of sludge and splitting the nitric acid feed. A reactor freeboard of 300% was again demonstrated to be necessary to control the foam. Also the use of an antifoam agent

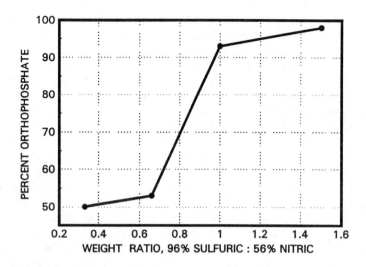

Figure 3. Conversion of elemental phosphorus to orthophosphates.

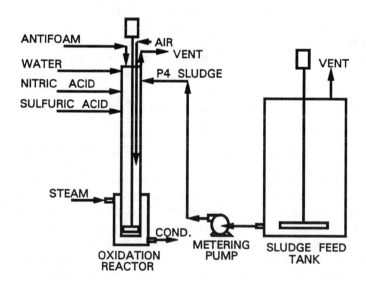

Figure 4. Bench-scale phosphorus sludge oxidation plant.

(sulfonated oleic acid) or increased agitator speed was shown to significantly aid foam control. During these bench-scale tests, no supplemental heat was needed during the 2 hour reaction time after the sludge addition was complete. After reaching a reaction temperature of 110-120 °C, the reactor cooled to 85-95 °C during this retention time without any noticeable reduction in the oxidation of the elemental phosphorus.

A portion of the oxidized sludge produced in the bench-scale oxidation facility was used as a feed to an ammoniator granulator to produce a plant nutrient product. These tests were made on a bench-scale granulation plant utilizing a 30 cm diameter by 30 cm long rotary drum granulator, followed by a 30 cm by 90 cm rotary dryer and a sizing screen. Ammonia was introduced in the granulator to neutralize and solidify the sludge and produce granules sized to approximately 2.4-3.4 mm. The granulation facility was operated at 6.8 kg/h and produced a product with a plant nutrient grade of 17-13-0-5S (17% nitrogen, 13% P_2O_5, and 5% sulfur). The product had an average pH of 5.0, and all the TCLP metals were analyzed and found to be below the regulatory levels. Table V shows a listing of these results from the granulation tests. Physical properties, storage, and greenhouse studies on the product all showed satisfactory results. The greenhouse tests showed the product to be at least as effective for plant growth as analogous rates of MAP (monoammonium phosphate) and no adverse effects of heavy metals were detected. Because of the success of the granulation study and the simplicity of the ammoniation/granulation process, scale up to a standard granulation pilot plant (0.5-1.0 t/h) should be routine.

Table IV. Bench-Scale Oxidation Results

Feed	
Sludge	
P_4, wt %	6.4
Total P, wt %	7.5
Acid Ratio, $H_2SO_4 : HNO_3$	1.1
Oxidation Product	
P_4, ppm	0.3 - 3.0
Total P, wt %	4.4 - 4.6
Orthophosphates, wt % of total P	97 - 99

These bench-scale tests successfully demonstrated the oxidation of elemental phosphorus containing sludge using mixed nitric and sulfuric acids and the conversion of the oxidation product into a usable granular nitrogen-phosphate plant nutrient product. These tests further verified the process dependence on the molar ratios of sulfuric acid to nitric acid, and nitric acid to elemental phosphorus as well as confirming the results of the earlier lab tests. Based on these test results, a patent application was submitted in March 1993. At this stage in the project, scale up of the oxidation and granulation processes to pilot plant scale was considered to be highly feasible and would provide an opportunity to complete the development of this technology while providing a technology demonstration with the processing of the approximately 130,000 liters of elemental phosphorus containing sludge currently in storage at TVA's Muscle Shoals facility.

Table V. Bench-Scale Granulation Results

Production Rate, kg/h	6.8
Product	
Total N, wt %	17.3
Total P_2O_5, wt %	13.9
Total K_2O, wt %	0.5
Total S, wt %	5.7
Total H_2O, wt %	0.9
pH	5.0
TCLP Metals	All Below Regulatory Levels

Pilot Plant Construction

Design and construction of a nitric acid oxidation pilot plant to process phosphorus sludge was started in early 1993. Completion and startup of the facility is scheduled for early fall 1993. The oxidation pilot plant is a scale up of approximately 20:1 over the bench-scale plant and will process up to 380 liters of sludge per batch. An existing granulation facility will be used for the pilot granulation work. Preferred operating parameters developed during the bench-scale tests will be used for startup of the pilot plant and are listed in Table VI.

Table VI. Preferred Operating Parameters

HNO_3 Feed, % of Stoichiometric	125 - 150
Molar Ratio of H_2SO_4 : HNO_3	1.1 - 1.3
Nitric Acid Concentration, wt %	56 -70
Sulfuric Acid Concentration, wt %	93 - 98
Phosphorus Sludge Concentration, wt % P_4	5 - 11
Reactor Temperature, $^{\circ}$C	100 - 115
Reaction Time, h	1.5 - 3.0

A basic flow diagram of the phosphorus sludge oxidation pilot plant is shown in Figure 5. Phosphorus sludge to be processed in the oxidation pilot plant will be removed from storage vessels into a large containment sump. In the sump the sludge will be recirculated through a chopper-type pump for size reduction and slurried for pumping to the pilot plant feed tanks. It should be noted that removal of any phosphorus sludge from its containment under water is an extremely hazardous operation. The sludge should be kept wet or preferably under water while it is being handled. Personnel working with the sludge should be dressed in fire protective suits and supplied with personal protective respirators designated for the copious amounts

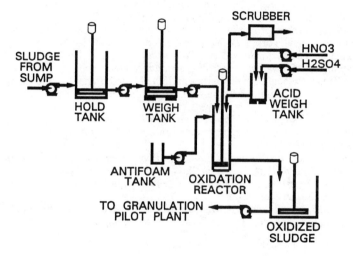

Figure 5. Phosphorus sludge oxidation pilot plant.

of P_2O_5 smoke that is produced by the oxidation of the phosphorus with air. Tanks will be available at the pilot plant to adjust the elemental phosphorus concentrations before feeding to the reactor. The reactor is a brick lined cylindrical vessel with a 0.76 m inside diameter and 3.35 m high. A natural gas thermal NO_x reducer was purchased for the pilot plant that is designed to reduce 90-95% of the NO_x produced to CO_2, H_2O, and N_2.

The first months of pilot plant operation will be dedicated to developing operating procedures and performing variable studies to further optimize the process technology. Beginning in early 1994, plans are to have the pilot plant operating on a 24-hour schedule with project completion targeted for August 1994.

Literature Cited

1.   Toy, A. D.; Walsh, E. N. *Phosphorus Chemistry in Everyday Living*; Second Edition; American Chemical Society: Washington, DC, 1987.
2.   Kirk-Othmer *Encyclopedia of Chemical Technology;* Third Edition; John Wiley and Sons: New York, New York, 1982, Vol. 17; pp 473-490.
3.   Hanna, J. *The Hazardous Waste Consultant*; May/June 1992, pp 1.9-1.12.
4.   Striplin, M. M. *Phosphorus from Abundant Nuclear Energy*; Symposium of Oak Ridge Associated Universities and The United States Atomic Energy Commission, Gatlinburg, Tennessee, August 26-29, 1968, TVA Publication No. CD 509.
5.   Farr, T. D. *Phosphorus Properties of the Element and Some of Its Compounds*; Chemical Engineering Report No. 8; Tennessee Valley Authority, Wilson Dam, Alabama, 1950.
6.   Sullivan, J. M. *Processing of Elemental Phosphorus Containing Residues to Fertilizer Intermediates by Nitric Acid Oxidation*; Progress Report of Chemical Research Department; CRD 92-9, Tennessee Valley Authority, Muscle Shoals Alabama, September 1992.
7.   Sullivan, J. M., Thrasher, R. D., Edwards, R. E. Recovery of Phosphates from Elemental Phosphorus Bearing Wastes. *U.S. Patent No. 5,275,639*, Jan 4, 1994.

RECEIVED March 14, 1995

Vitrification and Thermolysis

Chapter 9

Vitrification Technologies for the Treatment of Contaminated Soil

Laurel J. Staley

National Risk Management Research Laboratory, U.S. Environmental Protection Agency, 26 West Martin Luther King Drive, Cincinnati, OH 45268

Vitrification involves the melting and refreezing of soil to create a glass-like solid that entraps inorganic contamination thereby isolating it from the environment. The high temperatures required to melt soil also destroy organic contamination. Vitrification is thus capable of treating soil that is contaminated with both organic chemicals and metals. Six vitrification technologies have been studied thus far under the U.S. Environmental Protection Agency's Superfund Innovative Technology Evaluation (SITE) Program. This paper discusses the performance of each of these technologies.

Vitrification has been viewed as a potentially useful way to treat soil contaminated with both organic chemicals and metals. The high temperatures required to melt soil (2012-2642°F, 1100-1450°C) cause rapid volatilization and reaction of organic contamination.(1) The glassy solid products of vitrification processes are believed to isolate inorganic contamination from the environment. It is the entrapment of inorganic contamination that gives vitrification a potential advantage over incineration.

Vitrification used in the context of hazardous waste treatment involves the melting and re-freezing of soil to form a glass-like solid product. Glasses are characterized as homogeneous non-crystalline solids. Characteristics of glasses that may make them well suited for waste treatment are nonporousness and resistance to chemical reaction. These characteristics may make glasses stable for very long periods of time. In fact, volcanic glass, obsidian, can remain stable for millions of years. Given these characteristics, soil contaminants trapped within a glass-like solid may remain isolated from the environment far longer than with other methods of treatment. (Spense, R. Dr. Oak Ridge National Laboratory, personal communication, 1993)

In creating a glass-like solid product from soil, it is necessary to first melt the soil into a homogeneous molten phase. Calculations indicate that it takes approximately 1000 BTU/lb (0.3kWh/kg) to melt soil containing 20 % moisture. Most systems require more than this amount of energy to accommodate process heat losses.

There are a number of different ways to provide this energy. Fossil fuels and electrical energy can be used to melt soil directly. Alternatively, soil can be introduced into an already existing molten bath. Provided that the soil forms a homogeneous molten phase during processing and that it is cooled quickly enough to prevent the formation of crystals, it is likely that the resultant waste form is a
glass. (Spense, R. Dr. Oak Ridge National Laboratory, personal communication, 1993) This paper discusses six vitrification technologies under study by the EPA Superfund Innovative Technology Evaluation (SITE) Program.

Process Descriptions

The six technologies studied thus far under the SITE program use either fossil fuel combustion, direct electrical heating or plasma to provide the energy required for waste vitrification. All but one are *ex-situ* processes which require that the waste to be treated be excavated prior to treatment. SITE demonstrations have been completed on four of these technologies. The two remaining technologies are being studied under the SITE Emerging Technologies program. All six of these technologies are listed in Table I.

Fossil Fuel Systems. Three of the six systems described above derive the energy needed for vitrification from the combustion of fossil fuels.

Horsehead Resources Flame Reactor. The Flame Reactor developed by Horsehead Resources, is a 23 Million BTU/hr system that is designed to separate cadmium, lead and zinc and other volatile metals from contaminated soil. A schematic of the Flame Reactor is shown in Figure 1.(6) Feed, which has been dried to less than 5% moisture and ground to a PSD of 80% less than 200 mesh, enters the top of the Flame Reactor and is rapidly heated to 2000°C. There, metal contamination is reacted in the reducing gas of the reactor to form the pure form of the metal. Volatile metals such as cadmium, lead and zinc leave the reactor in the gas

Table I. Vitrification Technologies Under Evaluation by SITE

Technology	Power Source	Size lb/hr	Cost $/Ton	Notes
Retech Inc. Plasma Centrifugal Furnace (PCF)	Plasma Torch	120	757-2200[a]	SITE Demo 7/91
Horsehead Inc. Flame Reactor	Fossil Fuel	1790	208-932[b]	SITE Demo 3/91
Babcock & Wilcox Cyclone Furnace	Fossil Fuel	170	465-529[c]	SITE Demo 11/91
GeoSafe In-Situ Vitrification (ISV)	Electric Resistance Heating	8000-12000	250-600[d]	SITE Demo 4/94
Ferro Corp. Electric Furnace	Electricity	22[e]	N/A	Emerging Technol.
Vortec Corp. Oxidation and Vitrification	Fossil Fuel	806[f]	N/A	Emerging Technol.

[a](2)
[b](3)
[c](4)
[d](5)
[e](10)
[f]feedrate reported as tested. (Second Quarterly Report Assistance ID CR818194-01-0 December 31, 1991)

phase. They are separated from the slag produced by the non-volatile constituents in the feed in the Slag Separator. Downstream from the Slag Separator, the volatile metals are oxidized in the combustion chamber. Here they form metal oxides which, as solids, precipitate out of the gas stream to be collected. A demonstration of the Flame Reactor was completed in March 1991 at Horsehead Resources' research center in Monaca, Pennsylvania. Results showed that the recoveries of lead, cadmium and chromium can be as high as 80% or 99% depending on the method of calculation.(3) The oxide product is intended for resale and reuse. The slag produced, which contains nonvolatile metals, is intended to be inert and to pass the Toxicity Characteristic Leaching Procedure (TCLP) tests for metals.

Babcock & Wilcox Cyclone Furnace. The Cyclone Furnace developed by Babcock & Wilcox vitrifies soil by burning it in a high efficiency furnace developed originally to burn pulverized coal for electrical power generation. A schematic of the Cyclone Furnace is shown in Figure 2.(6) Combustion air introduced tangentially to the furnace, burns either coal or natural gas to create turbulent, high temperature conditions. Contaminated soil introduced at the center of the burner is rapidly heated past its melting point. The centrifugal flow of the combustion air forces the molten soil against the burner walls where it is further heated. Molten soil flows out of the furnace and is rapidly water quenched. A SITE demonstration of the Cyclone Furnace was completed in November 1991. The demonstration took place at Babcock & Wilcox's Research Facility in Alliance, Ohio. Test results showed that approximately 96% of the soil feed is retained in the slag which passed the TCLP test during the SITE demonstration of this process. The remaining 4% of the soil feed exits the cyclone furnace as fly ash. This material is trapped in a baghouse. Fly ash from this process may be enriched in volatile metal and may not pass the TCLP test. It may require subsequent treatment prior to final disposal.

Vortec Corporation Combustion and Melting System. The Vortec Combustion and Melting System (CMS) uses the high efficiency combustion of fossil fuels in cyclonic flow to rapidly heat and melt contaminated soil. A schematic of the CMS is provided in Figure 3. (6) Contaminated soil and, if needed, glass-forming additives, are preheated in a cyclonic flow precombustor. Combustion and melting are completed in a counter rotating vortex combustion unit. Melting is completed in a cyclone melter. Slag is separated from combustion gasses at the exhaust of the cyclone melter and is removed from the CMS and allowed to cool. Exhaust gasses are treated via electrostatic precipitation to remove particulate prior to being released to the atmosphere. The addition of glass-forming additives to the process enhances the products of the vitrified product. (7)

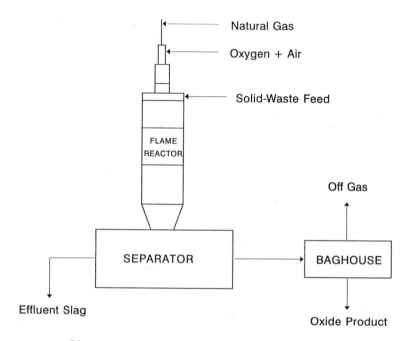

Figure 1   Horsehead Resources Flame Reactor

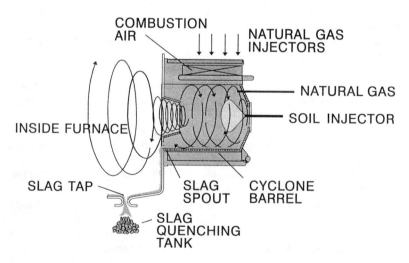

Figure 2   Babcock & Wilcox Cyclone Furnace Barrel

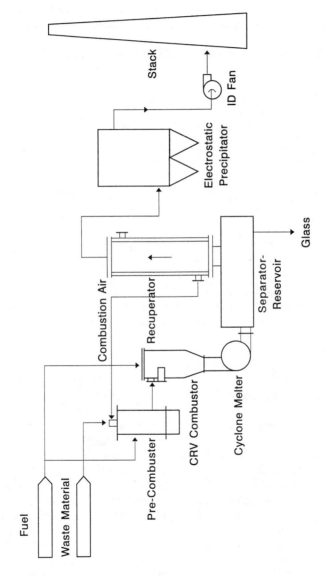

Figure 3 Vortec Oxidation and Vitrification Process

Horsehead Resources, Babcock & Wilcox and Vortec Corp. rely on the high efficiency combustion of either coal or natural gas to achieve vitrification temperatures. During their respective SITE demonstrations, each technology chose to burn natural gas. In no way were these technologies identical in operation, however. Differences in process design lead to significant differences in operating conditions as shown by Table II.

Table II. Summary of Operating Conditions for Fossil Fuel Systems

Technology	Fuel	Temperature °F, (°C)	Oxygen Level, %	Energy Use kWh/lb.
Horsehead Resources	Natural Gas	3632 (2000)	18[a]	3.09[b]
Vortec	Natural Gas	2000-3000 (1093-1648)	20[c]	1.27[d]
Babcock & Wilcox	Natural Gas	3000[e] (1648)	1	8.6

[a]The Flame Reactor used oxygen enrichment in its primary chamber at substoichiometric levels. The oxygen level reported here is for the secondary chamber where the volatilized metal is oxidized for proposes of precipitation. (9)
[b](3)
[c]Excess air, Oxygen level data are not available.
[d]Data for Case P, Reference 8, slurry of 30% water 70% dry solids. (8)
[e]Gas temperature. Slag temperature was 2360-2400°F.

Horsehead Resources achieves its very high process temperatures by burning natural gas using oxygen enrichment at substoichiometric conditions in the first stage of the process. That is, even though some of the combustion air was replaced with pure oxygen, the total amount of oxygen fed to the furnace was less than that required for complete combustion of the natural gas fuel. Because the combustion air/oxygen mixture has less inert nitrogen present to act as a heat sink, the combustion temperature is higher. The very hot oxygen deficient combustion gasses produced in the first part of the process reduce and vaporize volatile metals. The 18% oxygen level reported in Table II comes from the air used to oxidize these metals in the second stage of the Horsehead process. Because metal oxides are much less volatile than the pure form of the metal, they condense and are recovered in this stage of the process.

The unit consumption of energy is less for the Horsehead and Vortec systems than for the Babcock & Wilcox system. This may be an indication of economies of scale. Both the Horsehead and Vortec systems have feedrates that were roughly an order of magnitude higher than the Babcock & Wilcox system. As the feedrate of the Babcock & Wilcox Cyclone Furnace increases, the energy input per pound of feed may decrease somewhat.

Electric Technologies. GeoSafe and Ferro Corp. both use electric resistance heating to melt and vitrify soil. Ferro Corporation uses an *ex-situ* process to melt soil and adds glass-forming materials to facilitate the formation of glass from the molten soil. GeoSafe inserts electrodes directly into the ground and melts the soil through the direct application of electricity.

Ferro Corporation. The process developed by Ferro Corp. uses electrical resistance to melt a mixture of contaminated soil and glass-forming materials. A schematic of the Ferro Corporation process is shown in Figure 4.(6) Contaminated soil is fed into the process from the top. Contaminated soil is mixed with 33% by weight glass forming materials and is then heated past melting through the use of electrodes.(6) A bed of fresh soil feed lays on top of the molten glass/soil mixture as it is slowly incorporated into the melt from the bottom. This bed of fresh feed insulates the melt and traps organic contamination which is volatilized from the feed as it is heated by the molten bath below. As with other vitrification processes, inorganic contaminants are incorporated within the glass matrix upon refreezing. The molten glass/soil mixture at temperatures between 1500°C and 2000°C is discharged through the bottom of the reactor and fresh cold feed is added at the top. (10)

GeoSafe in-situ Vitrification. The *in-situ* vitrification process developed by GeoSafe, treats waste by heating it, in the ground through electrical resistance heating. A schematic of the process is provided in Figure 5.(6) The GeoSafe process uses four electrodes, inserted into the ground at varying depths. A small amount of glass frit is used to establish an electrical path between the electrodes at the start of melting. As the soil melts it becomes electrically conductive. Current then passes directly through the soil which accelerates heating. Soil gas passes up through the molten soil and is captured by a hood. Gas treatment consists of cooling, wet scrubbing, condensation, mist elimination and either HEPA filtration or activated carbon scrubbing to remove organic contamination. (11)

The operation of these two electrical systems is summarized in Table III below.

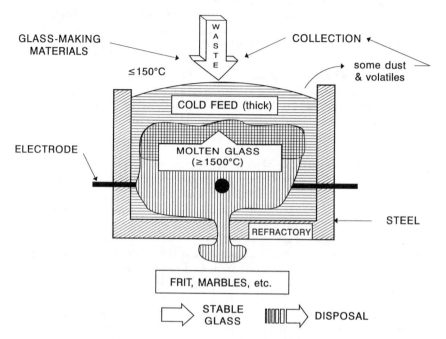

Figure 4   Ferro Corp.   Electric Furnace Vitrification

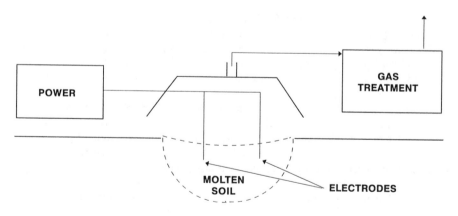

Figure 5   GeoSafe   In-Situ Vitrification

Table III. Summary of Operating Conditions for Electric Systems

Technology	Fuel	Temperature °F, (°C)	Oxygen Level,%	Energy Use kWh/lb.
GeoSafe	Electricity	2912-3632 (1600-2000)	N/A	0.83-1.1[a]
Ferro Corp	Electricity	3632 (2000)	21[b]	3.4[c]

[a](13,14)
[b]air above cold feed in the melter
[c](10)

The energy consumption per pound of feed is lower for these two technologies than for the fossil fuel systems discussed in the previous section. This may be because of the greater melt insulation provided by these two processes. The earth is a very good insulator. During ISV processing it has been observed that the 100°C isotherm is approximately one foot away from the melt which can be above 1600°C. (12) Since the large ISV melt is surrounded by the earth on all sides except for the very top, it is easy to understand why the melt might not lose much heat. The Ferro Corporation melter insulates the melt with a layer of contaminated soil. This too is a good insulator since the top of the melt is at approximately 100°C while the melt itself is at temperatures between 1500°C and 2000°C. (10)

Plasma Technology. Plasma systems have also been studied for their use in vitrifying soil. One plasma system has been demonstrated in the SITE program.

Retech Plasma Centrifugal Furnace. The Retech Plasma Centrifugal Furnace treats solid waste by incorporating it into a molten bath created by a transferred arc plasma torch. A schematic of the Centrifugal Furnace is provided in Figure 6.(6) The molten bath consists of contaminated soil which is retained along the walls of the centrifugal furnace by the rotating action of the reactor vessel. Upon initially entering the furnace, contaminated soil is incorporated into the molten bath. Organic contamination is volatilized and either reacts in the oxidizing atmosphere of the reactor, or flows out of the reactor and is oxidized in an afterburner downstream of the reactor. Exhaust gas from the reactor is treated in an aqueous scrubber prior to release to the atmosphere. Operating conditions for the Plasma Centrifugal Furnace are summarized in Table IV.

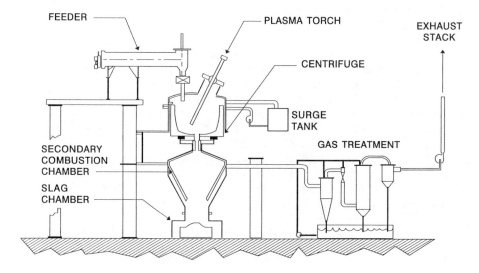

Figure 6 Retech Plasma Centrifugal Furnace

Table IV. Summary of Operating Conditions for the Plasma System

Technology	Fuel	Temperature °F, (°C)	Oxygen Level, %	Energy Use kWh/lb.
Retech PCF	Transferred Arc Plasma Torch	1800-2100 (982-1148)	11-14	6-8[a]

[a](15)

It should be noted that the temperatures listed in Table IV are for the gas phase above the soil. A SITE demonstration of the Plasma Centrifugal Furnace was completed in July 1991 at a Department of Energy research facility in Butte, Montana. Although soil temperatures were not measured directly during the SITE demonstration, it is believed that the soil at least momentarily achieved temperatures at or above its melting point since the soil directly beneath the torch was observed to glow and appeared molten. Although the soil was melted by the PCF to the point that it could flow out of the reactor and, upon hardening, form a rock-like hardened mass that passed TCLP, it was not the uniform glowing molten mass produced by the other processes discussed.

None of this is to say that the relatively low operating temperatures achieved by the PCF in any way impaired its ability to vitrify soil. Rather, it is to point out that in spite of the high temperatures achieved by plasma plumes, plasma vitrification systems do not necessarily generate temperatures significantly higher than conventional incineration. At 2100°C, the gas temperature produced by the PCF are lower than those for the other systems considered. The Babcock & Wilcox Cyclone Furnace, for example, achieved gas temperatures of 2800°F with a similar heat input per pound of feed. This indicates that less heat may be available for heating the soil in the PCF than in other systems. There may be several reasons for this. First, even though the energy supplied to this system is primarily used for creating the plasma, the soil does not pass directly through the plasma plume. Rather, it is heated by radiant energy from the plume and from some electrical conduction through the molten soil, which serves as a ground electrode in this systems. (16)

Secondly, a large amount of heat may have been lost through the PCF's water cooling circuits. Much of the primary and secondary chambers and the rotation mechanism is water cooled to prevent damage from thermal expansion.

Performance

Energy efficiency and cost factors are of secondary importance to performance. It does not matter how inexpensive something is if it does not work. As mentioned above, all of the technologies discussed in this paper have proven that they are capable of producing solid product that can pass TCLP. While that is certainly an essential characteristic of vitrification processes, it is not the only aspect of performance that is of importance. Vitrification technologies, like other technologies that are used to remediate Superfund sites, must comply with all Applicable Relevant and Appropriate Standards (ARARS).

Although ARARs for vitrification technologies will be determined by the appropriate regulatory bodies, one can speculate that performance standards and discharge limitations for process effluents may be considered ARARS. In addition, gaseous exhaust from these processes may need to meet Clean Air Act regulations. Liquid residues may need to meet Clean Water Act limitations if they are to be discharged on site. Vitrified material as well as any other solid residue produced by the site may need to meet the land disposal restrictions of the Hazardous and Solid Waste Act (HSWA). Further, since vitrification involves the thermal destruction of organic contamination, it is appropriate to compare the performance of vitrification processes against the Resource Conservation and Recovery Act (RCRA) incineration regulations. State and local standards will also need to be met for each site. Table V is a summary of some of the RCRA and HSWA performance standards compared against the performance of each of the six vitrification processes discussed in this paper.

Relatively little performance data is currently available on either the Vortec or Ferro processes. This is because they are currently being studied under the Emerging Technologies Program and have not completed field demonstrations of their technology.

DRE was not measured during the SITE demonstration of the Horsehead Flame Reactor. The waste used for this test consisted primarily of metals. It is very likely that, at the high temperatures achieved by this process, organic contamination would be thermally destroyed and DREs would exceed 99.99%. Generally, the vitrification processes discussed here were able to achieve at least 99.99% DRE. This is to be expected given the high operating temperatures achieved by each device.

Each of these technologies was able to produce solid material that passed the TCLP test. With the Flame Reactor it was necessary to add limestone to produce solid material that did not dissolve during the TCLP test.

With the exception of the Centrifugal Reactor, each of the six technologies discussed here released less than the 0.08grains/dscf particulate mandated by the RCRA hazardous waste incineration performance standards. Retech was not able to because of the performance of its air pollution control system. The design of this system can be readily changed and is not likely to affect the ability of the PCF to meet particulate emissions standards. (2)

Volatile metals were present in the exhaust gas from all of the processes tested. This is to be expected given the high temperatures achieved by each process. Nevertheless, metal emissions were not above regulatory limits with the exception of the Horsehead Flame Reactor. In this case the Flame Reactor exceeded standards set forth in the RCRA Boiler and Industrial Furnace Rules. It was decided during the Flame Reactor demonstration to use BIF standards instead of RCRA standards as ARARs. Some volatile metals were captured by the water present in scrubbers and quench systems. Those levels of metals were not high enough, however, to make it necessary to dispose of this residue as hazardous waste. (2,4)

While the exhaust gas from these processes contained very low levels of Carbon Monoxide and Total Unburned Hydrocarbon, levels of NO_x were elevated for some systems. Systems that have very high operating temperatures and use large amounts of combustion air seem to have the highest levels of NO_x. The presence of high levels of NO_x does not necessarily preclude the use of any of these technologies since NO_x abatement technology is available. Nevertheless, the need for NO_x abatement can raise the operating cost of these technologies.

Even if the residues produced by each process are less toxic than the original feed material, disposal still adds to the cost of operating each process. It is, therefore, valuable to know how much residual material is produced per pound of feed material treated. Estimates of this are provided in Table VI.

None of these technologies produces an extraordinarily large quantity of residue. The Babcock & Wilcox process produces a large amount of gas, but the ratio of gas to waste feed may decline as the process is scaled up.

Conclusions

A number of vitrification technologies have been shown to effectively treat soil contaminated with organic chemicals and metals. These technologies have been shown to entrap metals in a solid residue that passes TCLP. In addition, these technologies have been proven capable of thermally destroying organic contamination present within the soil feed. Solid, liquid and gaseous residues can be readily treated or disposed and are not produced in such a large quantity as to adversely affect the economics of using these processes.

Table V. Selected Performance Standards for Vitrification Technologies

Tech/ARAR	DRE	Particulate Emissions gr/dscf	Acid Gas lb/hr	Metals in the Exhaust Gas	TCLP Test Results glass/feed ppm	Other Exhaust Gas Constituents ppm
Incineration ARARs[a]	99.99%	0.08	4		Cd<1 Cr<5 Pb<5	
Retech[b]	99.9929-99.99965	0.37	0.017-0.00173	Pb, As	Zn 0.4/982[c]	500 NO_x <4 THC <2 CO
Babcock & Wilcox[d]	99.996-99.998	0.0008	N/A	Cd, Cr, Pb	Cd 0.12/49.9 Cr 0.22/2.64 Pb 0.31/97.3[e]	310-345 NO_x 5.9-18.2 THC 4.8-54.1 CO
GeoSafe	99.9999[f]	0.002[g]	N/A	Pb, Cd[h]		N/A
Vortec	N/A	0.3-3.3[i]	N/A	Pb, Cd	Cd 0.008/2.7 Cr 0.19/0.04 Pb 0.13/7.5[j]	200ppmNO_x

Ferro	N/A	N/A	N/A	N/A	Cu 0.5[k] Pb 0.1 Zn 0.3[l]	N/A
Horsehead	N/A	0.003-0.008	38.5-46.4[m]	N/A	As, Pb Cr[n] Cd 0.05/12.4 Pb 0.33/5.58	16-18.5ppmNO$_x$ 1-14ppmCO 1.6-0.91 THC

[a](17)
[b](15)
[c] Zinc was used as a substitute for TCLP metals. Compare against the regulatory level for Lead 0.05ppm
[d](4)
[e] Average values for all demo tests.
[f](18)
[g](11)
[h] no data available
[i] (Patten, J.; Lucas, P.; Santioanni, J.; Hnat, J.; *JAWMA*, in press)
[j](7)
[k] no TCLP data available for the feed
[l] Average values for all demo tests (10)
[m] BIF rules were ARARs for this demo. HCL emissions are limited to 0.72lb/hr
[n] These values exceeded the Tier II limits of the BIF rules.

Table VI. Treatment Residuals Produced per Pound of Feed
Treated

Technology lb/hr Residue lb/lb feed	Feed Rate Solids Residue lb/lb feed	Pounds Liquid Residue lb/lb feed	Pounds Gaesous	Pounds
Retech[a]	120	0.78	6.4	0.4
Babcock & Wilcox[b]	170	0.75	0.48	21
GeoSafe[c]	8000-12000	N/A	N/A	0.0145
Horsehead[d] 0.236 oxide	1790	0.323[e] slag	N/A	N/A
Vortec[f]	806	0.83	0.3	3.11
Ferro[g]	22	1.06[h]	N/A	N/A

[a](2,15)
[b](4,8)
[c](5)
[d](1)
[e](3)
[f](7)
[g](10)

For copies of reports which are available on these technologies and for more information on each of these technologies, contact the following individuals listed in Table VII

Table VII. EPA SITE Project Managers

Technology	Contact	Phone
Retech	Laurel Staley	(513) 569-7863
Horsehead Resources	Marta Richards Donald Oberacker	(513) 569-7692 (513) 569-7510
Babcock & Wilcox	Laurel Staley	(513) 569-7863
GeoSafe	Teri Richardson	(513) 569-7949
Vortec	Teri Richardson	(513) 569-7949
IGT	Teri Richardson	(513) 569-7949
Ferro Corp.	Randy Parker	(513) 569-7271

Literature Cited

1. Fitspatrick, V.P., Timmerman, C.L. and Buelt, J. *In Situ Vitrification - An Innovative Thermal Treatment Technology*; Proceedings of the Second International Conference on New Frontiers in Hazardous Waste Management EPA 600/9-87/018F   August **1987** PB 88-113360
2. *Retech Plasma Centrifugal Furnace: Applications Analysis Report* EPA540/A5-91/007 June **1992**
3. *Horsehead Resources Development Company, Inc. Flame Reactor Technology: Applications Analysis Report* EPA 540/A5-91/005 May **1992.**
4. *Babcock & Wilcox Cyclone Furnace: Applications Analysis Report* EPA540/AR-92/017   August **1992**
5. *In-Situ Vitrification for Permanent Treatment of Hazardous Waste* Presented at the Conference on Advances in Separations: A Focus on Electrotechnologies for Products and Waste April 11-12, **1989** Battelle, Columbus, Ohio
6. *The Superfund Innovative Technology Evaluation Program Technology Profiles* Fifth Edition EPA540/R-92/077 November **1992.**

7. Shearer, T., Hnat, J., Patten, J., Santioanni, J., and J. St. Clair, *Vitrification of Heavy Metal Contaminated Soils with Vortec Corporation's Combustion & Melting System (CMS)* Presented at the I&EC Special Symposium, American Chemical Society, Atlanta GA September 21-23, **1992**.

8. *Technical Evaluation Report: Babcock & Wilcox Cyclone Furnace* EPA540/SR-92/017.

9. *Technical Evaluation Report: Horsehead Resource Development Company Flame Reactor Technology* EPA/540/5-91/005   June **1992**

10. Spinosa, Emilio D. *Vitrification of Superfund Site Soils and Sludges* Presented at the Symposium on Environmental and Waste Management Issues in the Ceramic Industry.

11. Reimus, M.A.H. *Feasibility Testing of In Situ Vitrification of New Bedford Harbor Sediments*;   Ebasco Services Inc. Contract 2311113449 December **1988**

12. Koegler, S.S. *Disposal of Hazardous Wastes by In-Situ Vitrification*; Battelle Pacific Northwest Laboratory, Report for DOE under contract DE-AC06-76RLO 1830 (PNL-6281 UC-70)

13. Timmerman, C.L. *Feasibility Testing of In Situ Vitrification of Arnold Engineering Development Center Contaminated Soils*; ORNL/SUB/88-14384/1 PNL-6780 UC 602 March **1989**.

14. Hansen, J. and V. Fitspatrick "In- Situ Vitrification - Heat and Mobility are Combined for Soil Remediation", Hazmat World Feature Report, December **1989**. pp 30-34

15. *Technical Evaluation Report: Retech Plasma Centrifugal Furnace*; EPA540/5-91/007b **1992**

16. *Handbook: Vitrification Technologies for Treatment of Hazardous and Radioactive Waste*; EPA/625/R-92/002, May **1992**.

17. *Superfund Engineering Issue: Issues Affecting the Applicability and Success of Remedial/Removal Incineration Projects*;   EPA 540/2-91/004, February **1991**.

18. Timmerman, C.L. In *In Situ Vitrification of PCB-Contaminated Soils*; Report on Research Project 1263-24, Battelle Pacific Northwest Laboratories, Richland, WA, October **1986**.

RECEIVED March 14, 1995

Chapter 10

Iron-Enriched Basalt Waste Forms

G. A. Reimann, J. D. Grandy, and G. L. Anderson

Idaho National Engineering Laboratory, EG&G Idaho, Inc., P.O. Box 1625, Idaho Falls, ID 83415-2210

This paper reviews data on the iron-enriched basalt (IEB) waste form developed at the Idaho National Engineering Laboratory (INEL). IEB is a glass-ceramic waste form proposed for stabilization and immobilization of large volumes of low-level nuclear wastes for permanent disposal in an appropriate repository. IEB resembles natural basalt rock and is a product of melting a mixture of soil and retrieved wastes. Studies indicate that IEB will have a high tolerance for heterogeneous waste materials, including scrap metals, while maintaining the desired chemical and physical performance characteristics. Controlled cooling of molten IEB produces a glass-ceramic with good mechanical properties and high leach resistance. Production of IEB in a Joule-heated melter was difficult because of rapid electrode and refractory corrosion associated with high melt temperatures (1400 to >1600°C). Present studies include investigating the applicability of arc furnace technology to waste form production and determination of compositional limits for satisfactory waste form performance.

This paper summarizes results of research on the iron-enriched basalt waste form conducted at the Idaho National Engineering Laboratory during 1979-1982 (*1,2*) and from 1991 to present. This research investigated the applicability of thermal treatment to stabilize and immobilize between 227,000 and 340,000 m^3 (8 to 12 million ft^3) of wastes that were placed in "temporary" storage in the INEL's Radioactive Waste Management Complex (RWMC) since this facility was opened in 1952. (*3*) The figures given for the wastes include substantial quantities of overburden and underburden soils that may be contaminated and require treatment. Much of the material is low-level waste (LLW) contaminated with transuranic (TRU) elements from weapons-related work at the Rocky Flats Plant (RFP), and contains other substances defined as toxic and/or hazardous. Boxes and drums of these wastes were buried in pits and trenches between 1952 and 1970. After 1970 waste containers were stacked on asphalt pads, covered with soil for shielding, and protected with a membrane to

0097–6156/95/0607–0121$12.00/0

shed water. Some of the wastes have been stored beneath an air-supported canopy since 1976, which permits operations to proceed regardless of the weather.

The heterogeneous wastes stored at the RWMC include wood, paper, cloth, plastics, concrete, metals, chemical sludges, and contaminated soil. Many of the buried waste containers were ruptured by heavy equipment when leveling the soil cover during pit closure, while other containers were breached due to deterioration during long-term burial. The extremely variable nature of the mix of wastes complicates conversion into a suitable waste form. The waste stacked on pads is better characterized and more homogeneous, these containers are in better condition, and less entrained soil is present.

Thermal treatment is considered the best approach for most of the RWMC wastes as it would destroy the organic content, decompose the sludges, oxidize the metals, and generate a durable and versatile waste form. Interest in IEB as a waste form developed as a consequence of investigating the applicability of the slagging pyrolysis incinerator (SPI) for processing the RWMC wastes. (4) While applying the SPI to RWMC waste did not appear feasible, the slag produced by this furnace exhibited some very desirable characteristics. This aspect was pursued, and the IEB waste form evolved. (1) The IEB compositions described in this paper are the natural, unavoidable consequence of melting the mix of soil and waste encountered at the RWMC.

A variety of thermal processing options are adaptable to the conversion of the RWMC wastes into stable waste forms. Previous research at the INEL used a Joule-heated melter for pilot-scale melts. (5) Carbon arc melting (6) and in situ vitrification (ISV) (7) were investigated also. This paper will concentrate on the IEB waste form. Other waste forms and the various thermal processing technologies will be covered in other papers presented at this Symposium.

Natural Basalt

Natural basalt is a fine-grained, dark-colored extrusive igneous rock that covers large areas of the earth's surface. Natural basalt is composed primarily of calcic plagioclase and monoclinic pyroxene, with magnetite, olivine, and certain other minerals often present. It is mechanically durable and very resistant to weathering, and it seems logical that a manufactured waste form of similar composition would be similarly durable and leach-resistant. Natural basalts have been remelted and transformed into stable glass-ceramics that exhibit good chemical durability, high mechanical strength, and good resistance to abrasion. (8) Examples of microstructures of basalts from the U. S. Northwest are given in Figure 1.

Iron-Enriched Basalt

Soil from the RWMC area consists mostly of loess and clay-dominated sediments deposited in old stream channels of the Big Lost River. The composition of RWMC soil was obtained by analyzing a blend of samples taken from several locations. (9) The results of this analysis are given in Table I as "A-100." A similar soil, "SDA lake bed soil," was excavated outside the RWMC boundary and used as fill and to cover material stacked on storage pads. The lake bed soil has a slightly higher silica and

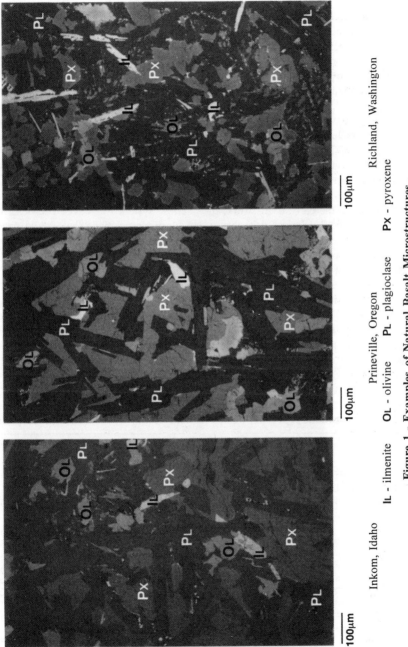

Inkom, Idaho

Prineville, Oregon

Richland, Washington

IL - ilmenite   OL - olivine   PL - plagioclase   Px - pyroxene

Figure 1 - Examples of Natural Basalt Microstructures

alumina content and was used as an ingredient in most of our experimental work. Soil is a major component in the formulations used to produce IEB.

Borosilicate glass (BSG) has been approved as a waste form for immobilization of high-level wastes, but converting several million cubic feet of LLW at the RWMC to the BSG composition would require a large amount of additives. While conversion to BSG would permit use of a low-temperature (1000-1200°C) melter, it would increase the waste form volume without improving the waste form properties, and it would require separation and removal of scrap metal that otherwise would accumulate unmelted in the low-temperature melter used for BSG. The temperatures used to produce IEB (above 1400°C, usually 1550-1650°C) could melt this scrap and hasten its oxidation while maintaining appropriate slag viscosity. The early studies showed that, in terms of leach resistance, the IEB waste form is at least as good as, and possibly better than, BSG. (*1,10*) Part of the present INEL technical mission is to generate sufficient data on IEB to justify its approval for use as a waste form.

A range of compositions is expected to result from the variety of RWMC wastes to be treated. In the early INEL work, the "Series" system was devised to describe these wastes. The "A-Series" describes waste form compositions that should result from mixing "**A**verage" TRU waste with "average" RWMC soil. The TRU wastes include sludges, combustibles, and noncombustibles (including metals). Because of the variable nature of wastes and soils, the likelihood of encountering a waste or soil sample that corresponds to the average is remote, so the compositions given are *reference* compositions only. The "A-100" in Table I describes the composition of 100% RWMC soil with no waste, while "A-0" describes the composition of average RWMC waste with no soil. The "A-40" composition, 60% waste and 40% soil, corresponds most closely to the composition defined as nominal IEB and is the basis for our studies. Compositional ranges of some natural basalts are included in Table I for comparison.

Table I. Compositions of RWMC Soil, Wastes, Nominal IEB, and Natural Basalt

	Oxide Compounds (wt%)								
	SiO_2	Al_2O_3	$FeO+Fe_2O_3$	CaO	MgO	Na_2O	K_2O	TiO_2	Misc.
RWMC Soil (A-100)	65.4	12.5	4.8	9.6	2.5	1.5	2.9	0.7	---
SDA Lake Bed Soil	69.9	13.2	4.7	4.1	1.9	1.5	3.3	0.8	0.5
A-0	38.0	7.4	34.5	8.3	4.6	4.8	2.4	---	---
H1-0	25.2	7.0	29.6	12.9	3.0	15.9	2.6	0.0	3.8
H2-0	32.0	4.7	27.9	22.9	6.3	3.3	2.2	0.2	0.4
S-0	35.4	4.8	40.9	14.7	3.5	0.3	0.4	0.0	0.0
P-0	20.6	9.3	22.8	39.8	7.6	0.0	0.0	0.0	0.0
N-0	0.7	0.3	36.1	0.1	0.0	37.3	25.5	0.0	0.0
M-0	2.5	5.1	75.1	1.0	0.0	0.0	0.0	0.0	16.3
IEB (A-40)	51.0	10.3	19.6	9.7	3.5	3.2	2.6	---	---
Natural Basalt (*11*)	45-55	13-17	11-14	7-10	4-10	2-4	1-5	1-5	---

With the accumulation of additional information describing the RWMC wastes, additional waste series were developed; (*12*) these are listed in the table also. Most of these wastes result from defense-related activities at the Rocky Flats Plant (RFP) and are stored in steel drums. H1-0 and H2-0 designations describe the RFP 741 and 742 sludges, respectively, with Portland Cement and Oil-Dri added to absorb free liquid. The S-0 series is organics (oils and solvents, RFP 743) immobilized by mixing with Micro-Cel E ($CaSiO_3$) and Oil-Dri. The P-0 series is chemical wastes (RFP 744) immobilized by mixing with Portland Cement and magnesia cement. The N-0 designation refers to evaporator salts (RFP 745) from solar drying of liquid wastes and consists mostly of sodium and potassium nitrates with limited amounts of other wastes and small amounts of Oil-Dri. The M-0 waste (RFP-480) is a variety of unleached scrap metals bagged in plastic and loaded into boxes or drums. In this work, waste form characterization assumes that all volatiles have evaporated, all combustibles have been burned, and only oxide residues remain.

Production of IEB

The slagging pyrolysis incinerator, mentioned previously, was devised primarily for treating municipal waste. This concept was the first considered for treatment of the RWMC waste because its 200,000 kg/day (220 tons/day) capacity could be large enough to treat the waste on a scale necessary to complete the task within a reasonable time frame. (*4*) The SPI could achieve high temperatures (1650°C), and its offgas could reach 1450°C. These temperatures are substantially higher than those normally encountered when incinerating municipal waste. Supplemental fuel requirements depended on the amount of combustibles in the waste. The byproducts were metal, a basaltic slag, and the offgas. However, the SPI did not perform well on municipal waste unless the feed was fairly uniform; system upsets were common when variable feed compositions were used. The SPI was a shaft-type furnace (see Figure 2), and thus inherently vulnerable to bridging in the descending column of waste and freezing of the bath due to "off normal" conditions caused by feed material variations. Such problems usually must be corrected manually, which may be difficult, and which cannot be tolerated when melting RWMC wastes that require remote operation of equipment.

However, the basaltic slag generated by the SPI furnace was of interest, as its composition was what would be expected when melting the RWMC waste and accompanying soil. The SPI slags were quenched in water to produce a frit. Examination of SPI slag samples disclosed bits of charcoal and beads of metal trapped in a glassy phase resembling obsidian, indicating that this slag was developed in a strongly reducing rather than in an oxidizing environment.

At that time (ca. 1979), experience with incineration of radioactive waste was lacking and few data were available regarding behavior of radionuclides in a basaltic waste form. Much of the early INEL research on IEB was performed in fractional-liter crucibles, as no equipment was available to do pilot-scale studies. Through an interagency agreement, the U. S. Bureau of Mines (USBM) conducted melting tests in a one-ton tilt-pour electric arc furnace at the Albany Research Center. (*6,13*) While operation was inefficient, with high refractory wear and rapid electrode consumption, the furnace was able to achieve bath temperatures of 1500 to 1700°C and to convert

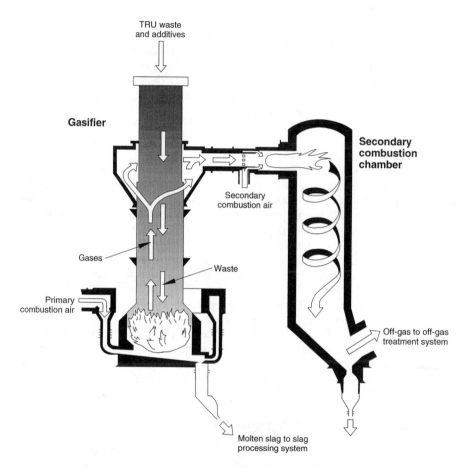

Figure 2 - Slagging Pyrolysis Incinerator System

various simulated RWMC wastes into basaltic melts that were poured into 208-L (55-gal) steel drums. This process appeared to have merit for conversion of mixed wastes into the IEB waste form; however, an arc furnace should not be used as an incinerator. The furnace had to be fed slowly when combustibles were charged, especially when significant polyethylene waste was present, to avoid generation of excessive soot and fumes that would produce unstable furnace operation.

In order to develop melter data at the INEL and produce IEB on a scale more applicable to production, an 80-liter joule-heated melter (JHM) was designed, built, and operated. (5) The decision to investigate the JHM concept was based on its successful application to large-volume production of BSG, and on the ability of Penberthy's electromelter to produce 90 kg experimental pours of simulated RWMC waste compositions. (1,14) Figure 3 shows a schematic of the INEL melter. Molybdenum electrodes and chromia-alumina refractories were used in this furnace to better endure the high temperatures. Electrode oxidation was troublesome and refractory attrition was high. Operating experience obtained with several versions of this furnace enabled performance to be improved. The most noteworthy design improvements were water-cooled refractory walls and incorporation of provisions to protect Mo electrodes from oxidation with a nitrogen blanket.

While additional design changes and adjustments to operating procedures would have further improved melter durability, the JHM experience encouraged the INEL to have another look at arc melters. The arc furnace used by the USBM for the work described previously has since been replaced with another unit that is more suitable for waste treatment. An American Society of Mechanical Engineers (ASME) test series for vitrification of incinerated municipal waste was performed in this furnace in 1992. (15) The INEL has a new interagency agreement with the USBM Albany Research Center to determine whether this arc furnace is suitable for treatment of the RWMC wastes on a larger scale. A test series was completed on simulated RWMC wastes in this furnace in July 1993 that entailed melting nearly 22,000 kg of material in five consecutive test days at feed rates of up to 700 kg/h and melt temperatures to 1850°C with no apparent equipment problems. (16)

Other pilot-scale arc melting approaches for radioactively-contaminated waste are under investigation, including the Pacific Northwest Laboratories/Massachusetts Institute of Technology (PNL/MIT) DC arc furnace and the Mountain States Energy/RETECH centrifugal reactor. Schematics of these three melting units are shown in Figure 4. Details of the progress of research efforts are contained in papers published in this Symposium.

Crystallization In IEB

Improved physical properties and leach resistance may be obtained from crystallized IEB, as compared to IEB that is mostly vitreous, especially if the radionuclides and/or toxic metals can be incorporated into the structures of crystalline phases. Slow cooling of IEB melts in the 1300 to 700°C range produces a mixture of crystals and a small amount of residual glass phase. In terms of leach resistance, crystalline structures are superior to glass of the same composition. Glass is metastable by definition and is more vulnerable to leaching than crystals of the same composition. (17)

The crystalline structures encountered in IEB are similar to those found in natural basalts, as may be observed in Figure 5; however, the redox conditions during melting,

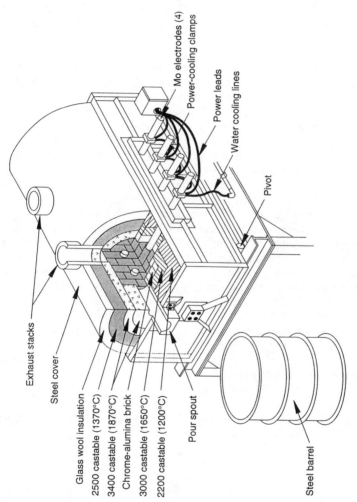

Figure 3 – INEL Joule-Heated Melter

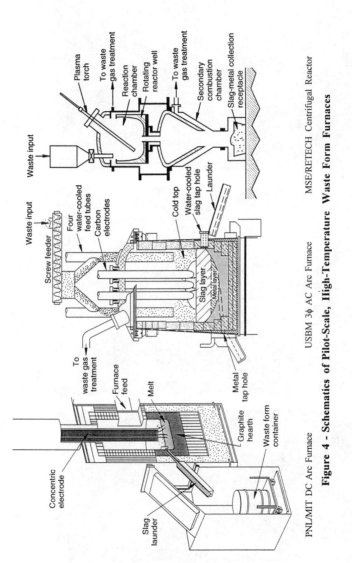

Figure 4 - Schematics of Pilot-Scale, High-Temperature Waste Form Furnaces

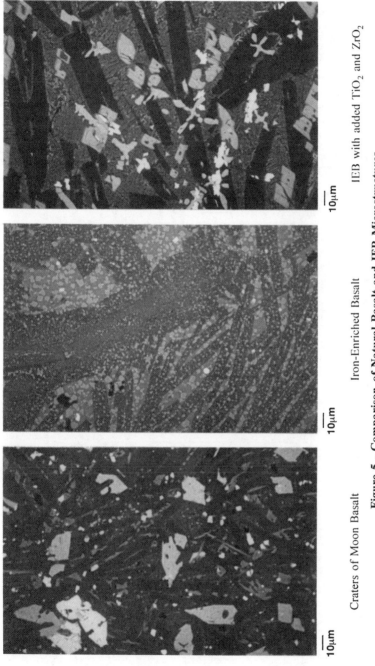

Craters of Moon Basalt   Iron-Enriched Basalt   IEB with added TiO$_2$ and ZrO$_2$

Figure 5 - Comparison of Natural Basalt and IEB Microstructures

the presence of the additional iron, and other oxides introduced during the course of waste assimilation may alter the composition sufficiently to affect the microstructure. The concentration at which the additives may have a deleterious effect on the desired properties of IEB has not been established and is a question that must be addressed. Up to this point, the major observed effects of compositional shifts have been on the quantities of crystalline phases present.

In the IEB compositions, iron spinels ($FeO \cdot Fe_2O_3$) are usually the first to crystallize from the liquid because of the amount of iron present, followed by augitic pyroxene [$Ca(MgFe^{+2}Al)(SiAl)_2O_6$] and calcic plagioclase. With further temperature decline, the plagioclase becomes more sodic, that is, the Ca-dominated plagioclase (anorthite, $CaAl_2Si_2O_8$) grades into an Na-dominated plagioclase (albite, $NaAlSi_3O_8$). One effect of the development of these crystals in IEB is to deplete the residual glass of iron, alkali, and alkali-earth oxides. The remaining glass thus becomes richer in silica and alumina and therefore becomes more leach resistant than the parent glass composition.

The ratio of FeO to Fe_2O_3 (or Fe^{+2} to Fe^{+3}) has a major influence on which crystals may nucleate and grow from the liquid. Beall and Rittler (8) reported that highly oxidized basaltic melts would crystallize into very durable glass-ceramics when $Fe^{+2}:Fe^{+3} < 1$. Redox conditions encountered during the course of melting, and the compositional variations within limits defining IEB, will alter significantly the composition of crystalline phases that develop, as well as the composition of the residual glass phase. Augite and/or olivine will form more readily when Fe^{+2} is abundant. Fe^{+3} becomes dominant in well-oxidized melts and anorthite may form before augite, or augite may not form at all due to the Fe^{+2} insufficiency.

The three melter types shown in Figure 4 operate with radically different redox environments, with the PNL/MIT furnace being the most reducing and the Retech centrifugal reactor being the most oxidizing. The USBM arc furnace will permit a range of redox conditions, depending on the amount of in-flow air permitted or whether the melt is oxygen-lanced, while the composition of the plasma gas in the Retech furnace may adjusted from reducing to pure oxygen. Feeding identical waste compositions into each furnace at the same rate and then subjecting them to identical cooling conditions may result in the formation of significantly different crystal species due to the different oxidation-reduction conditions.

Uranium and Transuranics in IEB

When small amounts of uranium or TRUs were added to IEB, these oxides dissolved into the melt as would any of the other oxide components. When these melts cooled, crystals of uranium or plutonium oxide were found in the residual glass phase. Without the presence of compounds to form suitable host crystals or solid solutions, the U and Pu oxides precipitated independently of the crystalline phases that normally form in cooling basalt, and were incorporated only if they became trapped within crystals that grew around them. Precipitated oxides of uranium and plutonium scattered throughout a glass phase would be more vulnerable to leaching than those incorporated into a leach-resistant crystalline phase. When sufficient zirconia (ZrO_2) was added to IEB, crystals of zirconia and/or zircon ($ZrSiO_4$) developed that incorporated uranium into their crystalline structures as solid solutions. (1) Zirconia and zircon are seldom encountered in natural basalts because ZrO_2 is insufficiently concentrated to form a crystalline phase.

Development of IEB4

Titania (TiO_2) is normally present in basalts in minor amounts. The RWMC soil contains about 0.7 wt% TiO_2. Adding TiO_2 to that already present in IEB to raise the total to about 5%, and adding a similar quantity of ZrO_2, will produce crystals of zirconolite ($CaZrTi_2O_7$) as the melt cools. (18) These crystals will be in addition to those that normally form in basalt. Natural zirconolite crystals have demonstrated a very high durability as well as a high capacity for U and Th. These properties should increase the leach resistance of the durable IEB waste form by immobilizing the TRUs within leach-resistant crystals.

IEB modified with additions of ZrO_2 and TiO_2 has been designated IEB4, in reference to the addition of these Group IVB elements. Efforts are presently underway to determine whether IEB4 will immobilize the TRUs in LLW in a manner analogous to the way the various formulations of Synroc immobilize TRUs in high-level waste. Zirconolite is formed with more difficulty in oxidized IEB4 melts with a high iron content because the titanium becomes tied up in the hematite and/or iron spinels. Pseudobrookite ($Ti_2Fe_2O_5$) may develop also, which has a small capacity (4 to 6%) for ZrO_2. Formation of these compounds depletes TiO_2 and ZrO_2 from the residual glass phase from which zirconolite must form.

Synroc requires hot pressing equipment while IEB4 develops zirconolite from a cooling basaltic liquid. Preliminary work on IEB4 was done using lanthanides as surrogates for the actinides. These studies are described in detail in another paper in this Symposium.

Conclusions

IEB is a waste form that is the unavoidable consequence of thermal treatment of toxic, hazardous, and/or radioactive INEL wastes that are mixed with soil. It closely resembles natural basalt in both composition and properties and thus would be a very durable material in which to stabilize and immobilize TRUs for time periods measured on a geologic scale. Adding small quantities (~5% each, or less) of TiO_2 and ZrO_2 enables a new suite of crystals to develop, so a very durable waste form may now contain TRUs within even more durable crystals.

The attributes of IEB have become well known; however, development of suitable industrial practices for treatment of large volumes of alpha LLW have lagged behind waste form development. Studies are now underway to develop and demonstrate techniques suitable for treatment of large waste quantities, and to determine if adjustments of these techniques are necessary in order to produce the desirable waste form properties.

Acknowledgments

This work was supported by the U.S. Department of Energy, Office of Environmental Restoration and Waste Management, under DOE Idaho Operations Office Contract DE-AC07-76ID01570.

Literature Cited

1. Flinn, J. E.; Henslee, S. P.; Kelsey, P. V.; Malik, R. K.; McCormack, M. D.; Owen, D. E.; Reimann, G. A.; Schuetz, S. T.; Seymour, W. C.; Tallman, R. L.; Welch, J. M. *Annual Report on TRU Waste Form Studies with Special Reference to Iron-Enriched Basalt: 1980*; EG&G-FM-5366, EG&G Idaho, Inc., Idaho Falls, ID, 1981.

2. Reimann, G. A.; Grandy, J. D.; Eddy, T. L.; Anderson, G. L. In *Nuclear and Hazardous Waste Management Spectrum '92 (Proc.)*; American Nuclear Society, Inc., La Grange Park, IL, 1992; Vol. 2, pp. 1083-1088.

3. Arrenholz D. A.; Knight, J. L. *A Brief Analysis and Description of Transuranic Wastes in the Subsurface Disposal Area of the Radioactive Waste Management Complex at the INEL*; EGG-WTD-9438, EG&G Idaho, Inc., Idaho Falls, ID, 1991.

4. Cox, N. D.; Nelson, D. C.; Burgus, W. F.; Richardson, L. S.; Hootman, H. E.; Stouky, R. J.; Nebeker, R. L.; Thompson, T. K. *Figure of Merit Analysis for TRU Waste Processing Facility at INEL*; TREE-1293, EG&G Idaho, Inc., Idaho Falls, ID, 1978.

5. Reimann, G. A.; Welch, J. M. *Electromelt Furnace Evaluation*; EGG-FM-5556, EG&G Idaho, Inc., Idaho Falls, ID, 1981.

6. Oden, L. L.; Johnson, E. A.; Sanker, P. E. In *Nuclear and Chemical Waste Management*; Pergamon Press Ltd., New York, NY, 1983; Vol. 4; pp. 239-244.

7. Callow, R. A.; Thompson, L. E.; Weidner, J. R.; Loehr, C. A.; McGrail, B. P.; Bates, S. O. *In-Situ Vitrification Application to Buried Waste: Final Report of Intermediate Field Tests at Idaho National Engineering Laboratory*; EGG-WTD-9807, EG&G Idaho, Inc., Idaho Falls, ID, 1991.

8. Beall, G. H.; Rittler, H. L. *Ceram. Bull.*, **1976**, *55*, pp. 579-582.

9. Reimann, G. A.; Grandy, J. D.; Eddy, T. L.; Anderson, G. L. *Summary of INEL Research on the Iron-Enriched Basalt Waste Form*; EGG-WTD-10056, EG&G Idaho, Inc., Idaho Falls, ID, 1992.

10. Hayward, P. J. In *Radioactive Waste Forms for the Future*; Lutze, W.; Ewing, R. C. Eds.; Elsevier Science Publishing Co., Inc., New York, NY, 1988; p. 471.

11. *Basalts*; Ragland, P. C.; Rogers, J. J. W.; Eds.; Van Nostrand Reinhold Co., Inc., New York, NY, 1984.

12. Bates, S. O. *Definition and Composition of Standard Wastestreams for Evaluation of Buried Waste Integrated Demonstration Treatment Technologies*; EGG-WTD-10660, EG&G Idaho, Inc., Idaho Falls, ID, 1993.

13. Nafziger, R. H.; Oden, L. L. In *Materials Research Symposium Proceedings,* Brookin, D. G., Ed.; Elsevier Publishing Co., Inc., New York, NY, 1983; Vol.15, pp. 639-646.

14. Flinn, J. E.; Kelsey, P. V.; Tallman, R. L.; Henslee, S. P.; Seymour, W. C. *Iron-Rich Basalt-Type Waste Forms for Transuranic and Low-Level Waste Containment: Evaluation of Electromelt Castings*; EGG-FM-5241, EG&G Idaho, Inc., Idaho Falls, ID, 1980.

15. O'Connor, W. K.; Oden, L. L.; Turner, P. C. In *Process Mineralogy XII*; Petruk, W.; Rule, A. R., Eds; The Minerals, Metals, & Materials Society, Warrendale, OH, 1994; pp. 17-37.

16. Oden, L. L.; O'Connor, W. K.; Turner, P. C.; Soelberg, N. R.; Anderson, G. L. *Baseline Tests for Arc Melter Vitrification of INEL Buried Wastes*; EGG-WTD-10981, EG&G Idaho Inc., Idaho Falls, ID, 1993; Vol. 1.
17. Roy, R. *Jour. Amer. Ceram. Soc.*, **1977**; *60*, p. 359.
18. Conley, J. G.; Kelsey, P. V.; Miley, D. V. In *Advances in Ceramics*; G. G. Wicks and W. A. Ross, Eds.; The American Ceramic Society, Columbus, OH, 1984; Vol. 8, pp. 302-309.

RECEIVED March 14, 1995

Chapter 11

Thermal Plasma Destruction of Hazardous Waste with Simultaneous Production of Valuable Coproducts

Q. Y. Han[1], Q. D. Zhuang[1], J. V. R. Heberlein[1,3], and W. Tormanen[2]

[1]Department of Mechanical Engineering, University of Minnesota, Minneapolis, MN 55455
[2]Minnesota Pollution Control Agency, 520 Lafayette North, St. Paul, MN 55155

Experiments for thermal plasma destruction of liquid hazardous wastes, such as PCBs, paint solvents, and cleaning agents, have been conducted in a counter-flow liquid injection plasma reactor. Efforts have been made to investigate the parameter space for optimal energy utilization while maintaining high destruction and removal efficiencies. Simultaneous production of diamond films while destroying liquid organic wastes has been demonstrated at the expense of lower destruction rates and higher specific energy requirements. Possibilities of increasing destruction rates and improving the specific energy requirements are discussed.

Thermal plasma pyrolysis of liquid hydrocarbons (chlorinated or non-chlorinated) is an attractive treatment process for ultimate disposal of chemical waste bearing these compounds (*1-3*). Due to its distinct features of high temperatures and high energy densities, thermal plasma pyrolysis may offer several advantages such as providing the potential for high destruction efficiencies, allowing the construction of compact, portable units with rapid start-up and shut-down capabilities, and independent control over the process chemistry. In several developments of plasma processes the major emphasis has been on compacting mixed solid and organic wastes, gasifying the volatile components and encapsulating the hazardous inorganic residues. Little work has been done exploiting the potential for generating valuable co-products.

In our previous experiments we have demonstrated the feasibility of destroying benzene with high destruction and removal efficiencies (DREs) but rather low destruction rates (*4, 5*). This feasibility study has now been extended to various liquid hazardous wastes including some paint solvents and polychlorinated biphenyls (PCBs). In a series of experiments we have investigated the parameter space for optimal energy utilization while maintaining DREs as required under the Resource Conservation and Recovery Act (RCRA) and the Toxic Substances Control Act (TSCA). We have also generated diamond films from these carbon-containing chemical compounds while trying to maintain high DREs. In this paper, we will report our experimental results on thermal plasma destruction of liquid organic wastes and simultaneous deposition of diamond films. Possibilities of increasing destruction rates and improving the specific energy requirements will be discussed.

---

[3]Corresponding author

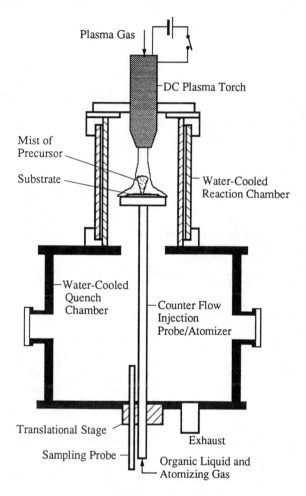

Figure 1. Schematic of the counter-flow liquid injection plasma reactor with substrate holder in position. (Reproduced with permission from ref. 8. Copyright 1992 Elsevier Science Publishers B.V..)

Experimental Apparatus and Method

In this laboratory, a counter-flow liquid injection plasma reactor has been shown to provide large reaction zones with enhanced mixing and relatively long reactant residence times (6,7). These features are considered to be of vital importance to enhance the economic viability of plasma waste destruction.

A schematic of the counter-flow liquid injection plasma reactor is shown in Figure 1. The reactor mainly consists of a DC plasma torch, a counter-flow liquid injection probe, a reaction chamber, and a quenching chamber. The reaction chamber is made of an inner quartz tube and an outer Pyrex glass tube with a water cooling channel in between. The counter-flow liquid injection probe consists of four concentric stainless steel tubes. Liquid toxic waste is fed through the innermost tube. Atomizing gas is introduced through the channel around the center tube. Next to it are the cooling water channels. The liquid injection probe is centered with respect to the plasma jet by a x-y-z translation stage. More detailed descriptions about the waste destruction system have been given in our previous papers (4, 5). Note that a water-cooled substrate holder has been added to the tip of the liquid injection tube to hold a substrate for diamond nucleation and growth (8). An aligned hole in the substrate makes the probe still function as an atomizer for liquid organic precursors.

During experiments, liquid organics are fed into the plasma reactor by a precision metered pump via the liquid injection probe. Atomizing gases (O_2, H_2 etc.) are used to atomize the liquid jet at the exit of the injection nozzle and jet of liquid droplets penetrates into the low density plasma plume. This flow configuration leads to the formation of a zone of recirculating flow between the jet and the reactor wall, resulting in good mixing and long residence times (> 100 ms) in regions of relatively high temperature (> 4000 K), and assuring complete vaporization and pyrolysis of organic compounds. The atomizing gas also serves as a reactive gas, playing an important role for preventing the regeneration of the original toxins. Samples of the product gases are collected for chemical analysis, via the sampling probe by means of two methods: one uses a 50 ml pre-evacuated glass flask; the other uses a sampling unit consisting of charcoal sampling tubes and a SKC flow sample pump. The samples collected in glass flasks are analyzed using a mass spectrometer which has a 10 parts-per-billion (ppb) level of detection ability for a wide range of masses. The samples collected in charcoal tubes are analyzed by the Twin City Testing Corporation, using a Hewlett-Packard 5890A gas chromatograph equipped with a flame ionization detector. According to the chemical analysis results, the destruction and removal efficiency (DRE) for each initial organic compound is determined as:

$$DRE = (\dot{M}_{in} - \dot{M}_{out}) / \dot{M}_{in} \bullet 100\%$$

where $\dot{M}_{in}$ is the mass feed rate of the compound, measured by a liquid flowmeter, and $\dot{M}_{out}$ the mass emission rate of the compound residual. The mass emission rate is determined as the total mass detected in the off-gas divided by the time during which the off-gas samples are collected.

Results and Discussion

1) Investigation on upper limit of waste destruction rates. Some experiments have been first conducted to further investigate the feasibility of complete destruction of various liquid hazardous wastes including hydrocarbon compounds and chlorinated hydrocarbon compounds. The hydrocarbon compounds used include acetone (C_3H_6O), benzene (C_6H_6), methyl ethyl ketone (MEK, C_4H_8O) and lacquer thinner, which are typical in paint solvents and cleaning agents. In these experiments, the plasma gas (Ar) flow rate has been 24 l/min. Oxygen has been used as the atomizing gas with flow rates ranging from 12 to 16 l/min. The power input has been in the range of 12.5 to 16.5 kW. Under these conditions, for liquid waste feed rates less than 5 ml/min, a chemical analysis of the exhaust reveals that carbon dioxide, water, carbon monoxide, hydrogen, and oxygen are the main products of these processes. No residuals of these hydrocarbon compounds have been found in the off-gas samples. The DREs for these processes are better than 99.9999%. Other experiments have been conducted with carbon tetrachloride (CCl_4) and polychlorinated biphenyls (PCB 1242, $C_{12}H_7Cl_3$). With chlorine in the processes, the main product gases are, carbon, carbon monoxide, water, hydrogen, and hydrogen chloride.

For the PCB destruction experiments considerable effort had to be spent for obtaining the permits and setting up the procedures and quality control measures. With the permit for carrying out bench scale PCB test burns approved by U.S. EPA in September, 1992, and the permit for air emission approved by Minnesota Pollution Control Agency in March, 1993, experiments have been conducted for testing the reactor's capability of destroying PCBs. In a series of experiments, Diala AX oil containing 53,000 ppm PCB (Aroclor 1242) has been injected into the plasma reactor for destruction. The flow rates have been up to 10 ml/min, which is the maximum amount allowed by the permit issued by EPA. Oxygen has been used as atomizing gas, with flow rates ranging from 10 to 15 l/min, to carry the PCB compound into the plasma plume. The off-gas samples have been analyzed by the Twin City Testing Corporation, using the method listed as NIOSH 5503. No PCB residuals have been detected in any of these off-gas samples, indicating DREs exceeding 99.999% for all of these PCB experiments. The experimental parameters and DREs of these processes are listed in Table I.

Table I Experimental Parameters and Results

Test run	# 1	# 2	#3
Sampling time	40 min.	40 min.	50 min.
Plasma torch power	35 V x 400 A	35 V x 400 A	32 V x 400 A
Plasma gas (Ar)	24 l/min	24 l/min	24 l/min
Atomizing gas (O_2)	10 l/min	15 l/min	15 l/min
PCB feeding rate	5.0 ml/min	5.0 ml/min	10 ml/min
PCB emission rate*	< 0.875 µg/min	< 1.0 µg/min	< 0.8 µg/min
DRE*	>99.99976%	>99.99973%	>99.99989%

* values reported are based on the analytical detection limit
SOURCE: Reprinted with permission from ref. 9. Copyright 1993.

Efforts have been also made in quantifying the upper limits of the benzene destruction rate and carbon tetrachloride destruction rate for the present reactor configuration at certain operation conditions. A destruction and removal efficiency (DRE) of 99.99% was selected as the lowest tolerable limit. Both oxygen and hydrogen have been used as atomizing gas with various flow rates. For benzene (C_6H_6), a destruction rate of 24 ml/min has been reached, with 16.0 liter/min O_2 as the atomizing gas and 12.6 kW power input to the plasma torch. This results in a specific energy requirement of 10 kWh/kg for benzene destruction. It should be noted that the specific energy requirement for benzene destruction has been substantially reduced as compared to the previously reported 138 kWh/kg. It is believed that this number can be further brought down by increasing the oxygen flow, though a trade-off may have to be made between reduced energy cost and increased oxygen cost. For carbon tetrachloride (CCl_4), the highest destruction rate has been 21 ml/min, with 1.5 liter/min and 16.0 liter/min hydrogen fed into the torch and the atomizing probe, respectively. The power input is 15.8 kW, thus leaving a specific energy requirement of 7.9 kWh/kg for the destruction of carbon tetrachloride. It should be noted that these values are only a factor of 6 to 10 higher than the specific energy requirements achieved by Westinghouse in a commercial scale 1 MW reactor.

2) Investigation on production of valuable co-product: diamond film. One of the objectives of this study has been to investigate the possibility of simultaneously obtaining valuable co-products while destroying organic wastes. Experiments have been carried out to generate diamond with all the above-mentioned organic compounds, including lacquer thinner and PCB contaminated oil. In addition, chloroform ($CHCl_3$), dichloromethane (CH_2Cl_2), trichloroethylene (C_2HCl_3), and 1,2-dichloroethane (C_2HCl_4) have been used as precursors for diamond deposition. Diamond films have been successfully deposited from all of these precursors. Some SEM micrographs of typical diamond films are shown in Figure 2. These diamond films are made from acetone, benzene, dichloromethane, carbon tetrachloride, lacquer thinner and PCB contaminated oil. It is seen from the SEM micrographs that the diamond crystals at the growth surface are well-faceted.

While diamond films have been successfully deposited from these precursors, efforts have been made to maintain relatively high DREs. However, the destruction rates have to be kept low (up to several milliliter per minute), since higher feeding rates have led to the deposition of graphite on the substrate and the formation of carbon powder in the reactor. Compared with the experimental results discussed earlier in this paper where 24 ml/min benzene and 21 ml/min carbon tetrachloride have been destroyed with satisfactory DREs, it is obvious that a gap exists between the optimal waste destruction rate and the waste destruction rate with diamond growth. In order to guide the diamond deposition experiments and interpret the experimental results, a C-H-O phase diagram for diamond deposition (*10*) has been adapted as shown in Figure. 3. The shaded area is a "diamond growth domain" where diamond growth has been reported using various hydrocarbon sources such as CH_4, C_2H_2, CO_2, etc.. Above the diamond-growth belt is the non-diamond carbon growth region, and the area below represents the no-growth region (neither carbon nor diamond growth). This phase diagram has been used for choosing the operating conditions in our experiments.

Data points from experimental results with benzene test runs have been put into the C-H-O phase diagram. Our data confirm the results of other investigators with respect to diamond growth from other hydrocarbon/gas mixtures. The results from tests No. 7 through No. 16 fall all on the dashed line (benzene line) and represent the data from a given amount of oxygen and different amounts of benzene. It is interesting to see that for a certain oxygen flow rate, increasing benzene input may bring the data points into the diamond-growth region. Therefore efforts have been made to deposit diamond film with benzene using only O_2 as carrier gas, following the prediction by the C-H-O phase diagram. However, so far diamond growth has not been observed in these experiments. This is probably due to the strong oxidization environment above the central portion of the substrate, thus preventing diamond nucleation. In several experiments, it has been found that Mo substrates are oxidized, thus they are no longer suitable for diamond nucleation and growth. To solve this problem, the possibilities of injecting different oxygen/hydrogen mixtures through the probe will have to be investigated further. In other words, a local reducing environment above the central portion of the substrate needs to be sustained so that diamond can nucleate and grow on the substrate.

Similar to the C-H-O phase diagram, we have constructed a C-H-Cl phase diagram for the study of producing diamond from chlorinated hydrocarbon

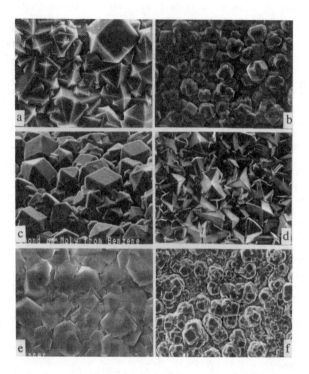

Figure 2. SEM micrographs of diamond films deposited from: (a) Acetone,
(b) Dichloromethane, (c) Benzene, (d) Carbon tetrachloride, (e) Lacquer thinner,
(f) PCB compound.

compounds (see Figure 4). Data points have been collected from experiments using carbon tetrachloride (CCl_4), chloroform ($CHCl_3$), dichloromethane (CH_2Cl_2), trichloroethylene (C_2HCl_3), and 1,2-dichloroethane (C_2HCl_4). In these experiments, power inputs to the plasma torch are in the range of 13 kW to 16.5 kW. The argon flow rate has been kept at 24 l/min, and 1.0 ~ 2.0 l/min hydrogen is added to the plasma gas to generate atomic hydrogen in the plasma plume. As atomizing gas, hydrogen has been used to carry and spray liquid precursors into the hot zone of the plasma jet. Diamond has been generated from these precursors with liquid flow rates up to 4.2 ml/min.

It is apparent that the combination of a high rate waste pyrolysis process with a diamond synthesis process has to overcome two sets of conflicting requirements: (1) the need for an oxidizing environment for the most efficient waste treatment versus the need for a reducing environment for the diamond deposition, and (2) the need for high flow rates of the waste stream in the treatment process versus the limitation of the fraction of the waste flow rate compared to the hydrogen flow in the case of diamond deposition. With the present reactor configuration, it is difficult to find a window where diamond is produced while at the same time wastes are being destroyed with sufficiently high feeding rates. In order to reduce the specific energy requirement of waste destruction while maintaining high destruction and removal efficiencies, a redesign of the reactor has been proposed. In the new design, a thermally insulated wall replaces the water-cooled wall of the reactor, and a second oxidation chamber is introduced. The use of the thermally insulated wall will substantially increase the reactor wall temperature, reduce the heat loses, and result in an expanded reaction zone for waste decomposition. The introduction of the second oxidation chamber will generate a flame-contact combustion process which can destroy any remaining toxic waste from the main reaction chamber and prevent recombination of toxins. With this two-stage reactor, the separation of the reducing environment for diamond growth and the oxidizing environment for most efficient waste treatment will be accomplished. However, the limitation of the waste flow compared to the hydrogen flow remains. This requirement of high hydrogen flow rates will limit the practicality of the combined process - waste pyrolysis with diamond deposition - if large scale incineration processes are an option. Use of some hazardous wastes as reactants in a diamond reactor may be more attractive, since we have demonstrated high diamond growth rates and high carbon conversion efficiencies with liquid reactants. The practicality then depends on the cost of (or credit for) the reactant feed stock versus that of off-gas monitoring. One might consider the additional case where on-site treatment of a small hazardous waste stream is preferable to transport. Under these conditions, a reactor which provides diamond as a co-product at the expense of adding hydrogen may prove attractive.

The experiments described here are all limited to the use of liquid hazardous wastes. It may be possible to extend the process to destruction of waste in the form of slurries. However, the described reactor is not suitable for treatment of wastes in which containment of major portions of the waste stream is required, e.g. wastes containing radioactive materials (mixed wastes), because it has no provision for collecting hazardous solid residues.

Conclusions and Summary

We have demonstrated that the counter-flow liquid injection plasma reactor is suitable for destroying a variety of liquid hazardous wastes with the destruction and removal efficiencies meeting the requirements mandated by the RCRA. Experiments performed in the present plasma reactor have shown its capability for destroying thermally stable organic compounds including PCBs. The destruction rates and specific energy requirements for benzene and carbon tetrachloride have been

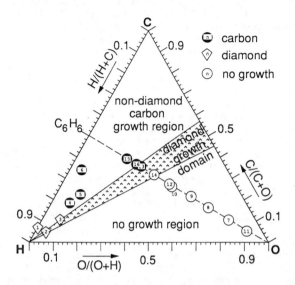

Figure 3. The C-H-O phase diagram for waste destruction/diamond deposition process using benzene as precursor. (Adapted from ref. 10.)

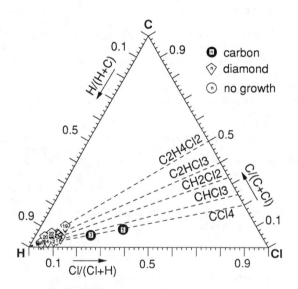

Figure 4. The C-H-Cl phase diagram for waste destruction/diamond deposition process using chlorinated hydrocarbons as precursors.

substantially improved, compared with the preliminary results from previous experiments. The present small scale reactor has not been optimized with regard to obtaining the same specific energy requirements as large scale installations, but rather to demonstrate the rapid and efficient destruction of limited amounts of liquid wastes. Diamond synthesis from the hazardous wastes, including hydrocarbons, chlorinated hydrocarbons and PCB contaminated oil, has been demonstrated, but the necessary process adjustments have led to a reduction in the waste destruction rate. Future efforts should be directed to improve the destruction rates in the combined process through previously mentioned reactor modifications.

Acknowledgments

The authors would like to acknowledge Dr. Pfender for his continuous advice and personal involvement in this research project. Mr. Y. Ma and Mr. B. Forsyth are acknowledged for performing the mass spectrometric analyses. This research work has been supported by a grant from the Blandin Foundation.

Literature Cited

1. Freeman, H.M.; Olexsey, R.A.; Oberacker, D.A.; Mournighan, R.E. *Journal of Hazardous Materials,* **1987**, *14*, p. 103.
2. Lee, C.C.; Huffman, G.L. *Environmental Progress*, **1989**, *8(3)*, p. 190.
3. Staley, L.J. *Proc. 1st INEL Plasma Waste Workshop*, Idaho Falls, ID, **1991**.
4. Han, Q.Y.; Heberlein, J.; Pfender, E. In *Heat Transfer in Thermal Plasma Processing;* Etemadi, K.; Mostaghimi, J., Eds.; ASME-HTD-Vol. 161; American Society of Mechanical Engineers: New York, NY, **1991**; pp. 53 - 56.
5. Han, Q.Y.; Heberlein, J.; Pfender, E. *Journal of Materials Synthesis and Processing*, **1993**, *1(1)*, p. 25.
6. Paik, S.H.; Chen, X.; Kong, P.; Pfender, E. *J. Plasma Chem. Plasma Process.* **1991**, *11*, p. 229.
7. Or, T.W.; Kong, P.C.; Pfender, E. *J. Plasma Chem. Plasma Process.* **1991**, *12*, p. 189.
8. Pfender, E.; Han, Q.Y.; Or, T.W.; Lu, Z.P.; Heberlein, J. *Diamond and Related materials*, **1992**, *1*, p. 127.
9. Han, Q.Y.; Heberlein, J.; Pfender, E. In *Proc. of the 11th International Symposium on Plasma Chemistry;* Harry, J., Ed.; IUPAC: Loughborough, England, **1993**; Vol. 2, pp. 758-63.
10. Bachmann, P.K.; Leers, D.; Lydtin, H. *Diamond and Related Materials*, **1991**, *1 (1)*, p. 1.

RECEIVED June 21, 1995

Chapter 12

Plasma-Assisted Cleaning of Flue Gas from a Sooting Combustion
Case of Organic Nitrates

A. Czernichowski[1], Ph. Labbé[2], F. Laval[2], and H. Lesueur[1,3]

[1]Université d'Orléans, 45067 Orléans Cedex 02, France
[2]Commissariat à l'Energie Atomique, 37260 Monts, France

Multi-electrode reactors based on the principle of Gliding Arcs (GlidArc) are used to clean-up highly soot and NOx charged Flue-Gases from an open burning of the Nitro-Benzene (laboratory scale) and Tri-Nitro-Toluene (pilot scale). High cleaning efficiency was achieved at the Flue-Gas flow-rate up to 200 m³/h and the electric power of less than 10 kW: almost complete disappearance of soot (and pyrolytic hazardous products adsorbed on it), and substantial lowering of the NOx and CO concentrations. Other applications of these new plasma reactors for large gas-flow processing are proposed.

Approximately 200,000 tons of obsolete conventional explosives and propellants are stored throughout the United States, and quite the same amount in Europe. The military's mounting stores of these "energetics" become less stable with age. Aggravating the disposal problem is the huge inventory of energetics that is accumulating as a consequence of military down-sizing. As storage space is rapidly filled up, the traditional disposal method, open burning, will be less of an option in the future. The equipment to conduct such an operation is rather simple and the destruction cost is minimum, but huge quantities of combustion products have been released into the air. Therefore, a ban on open burning proposed by the U.S. EPA that may take effect in 1995 forces the military to improve the combustion, to clean-up the flue-gas or to search for other disposal methods. The same restrictions are foreseen everywhere. The simplest way to eliminate an explosive material seems to be its detonation in open atmosphere, under the water, or under the ground, but a relatively large part of the initial toxic material becomes dispersed into the soil and the flue-gas problems remain the same.

While the organic waste is destroyed during an incineration, other hazardous material may be formed. Usually, an open burning of fossil fuels or waste organics leads to the so-called Products of Incomplete Combustion (PICs), which are formed due to incomplete conversion of organics to CO_2. Even in the case of some compounds

[3]Current address: BRGM, Avenue de Concyr, 45060 Orléans Cedex 2, France

0097–6156/95/0607–0144$12.00/0
© 1995 American Chemical Society

that are relatively rich in oxidizing -NO2 groups (like explosives), open burning may lead to the presence in fumes of toxic PICs like Polycyclic Aromatic Hydrocarbons (PAHs), Volatile Organic Compounds (VOCs), CO, and residual NOx. Of the main concern are PAHs, many of which have been identified to be carcinogenic and mutagenic (*1-3*). In flame environment, PAHs are adsorbed on the active surface of the soot. Soot particulates may contain up to 50 % of primary compounds or PICs. Due to their small size, soot particulates may be transported into the human lungs where they remain in contact with the lung tissue for a long time.

Soot deposits can also have adverse effects on heat transfer characteristics or combustion behavior, which can cause efficiency or failure problems in various systems as rocket or diesel engines, or waste incinerators.

PAH and Soot Formation

Over the past few years, there has been great interest in mechanism of PAH and soot formation in combustion (4). All available data show a central role of the first aromatic rings formation mechanism even in the flames of aliphatic fuels. Aromatic compounds in a fuel enhance the PAH and soot formation. This is the case of the combustion of such substances as Benzene (highly toxic), Toluene, Xylene, their mixture (BTX), and its derivatives like Tri-Nitro-Toluene (TNT) $CH_3 \cdot C_6H_2 \cdot (NO_2)_3$, the main subject of this work.

According to (*5*), there is no direct limit to total aromatics content in gasoline in Europe. The U.S. aromatics limits will be 28 vol. % in 1994, and 25 vol. % in 1996. The European Standards Organization seeks however a 10 vol. % limit in the BTX content in motor diesel fuel because of the aromatics role in the formation of toxic particulates in diesel exhaust. The solid phase or so-called "soot" contains fine particulates and is at a level much higher than that emitted from a gasoline-powered engine. The soot, which is formed as a consequence of the way in which the diesel fuel burns inside the engine, consists of carbon and adsorbed materials such as unburnt HydroCarbons (HCs), sulfates and metals.

The detailed formation and growth mechanism of PAH species (up to 11 condensed rings of the molecular mass up to 448 were already identified) and ultimately soot formation and deposits are very topical combustion problems (*6,7*). This can be described, for example, by the simplified Frenklach theory of the first aromatic ring formation (*6*). It is interesting to underline, that PAH removal by oxidation occurs predominantly through molecular oxygen attack on aromatic radicals $C_6H_5 \cdot$ (which propagates molecular growth and cyclization of PAHs). Due to similar reactions of O_2 with other radicals, the concentration of these critical intermediates decreases, reducing the formation rate of the first aromatic ring. But at the same time, the presence of O_2 in the burning mixture has a promoting effect on the aromatics formation resulting in an accelerated chain branching that leads to enhanced fuel pyrolysis and thus increases production of critical intermediates. Probably it is the reason why the aromatics are so hard to be completely burnt, even with a large excess of the air in a diesel motor or during an open burning of organic compounds.

Soot Emission Control

The control of combustion wastes is becoming more and more compelling towards operators. The next world's rules will be probably more and more applied to ordinary wastes as well as to the specific ones resulting from the destruction of the explosive materials.

During the last few years, several plants have been tested or proposed for the disposal of explosive and "propergol" wasters or materials from weapons dismantling, and eventually for their chemical reprocessing based on the following processes :
- steam-assisted melting of the TNT (and some other specific explosives) in order to recover (and then to reuse) the explosive matter;
- controlled incineration of explosives in stationary or rotative furnaces with a flue-gas (FG) cleaning (but just washing the FG does not solve the disposal problem of the recovered soot);
- molten salt oxidation;
- use of a super-critical fluid for selective material extraction;
- chemical neutralization;
- reusing a diluted explosive as a fuel;
- bio-decomposition by bacteria or filamentary fungi.

However, because of substantial investment costs and relatively small added value of products in the weapons dismantling industry, projects have not been carried out. Now, looking for solutions involving a small scale and less expensive installations becomes more and more obvious. Such units could be transported to sites to treat wastes that cannot be shipped off the sites for treatment.

Particular attention must be given to the technical control of the exhaust emission of particulate matter from diesel vehicles which has become a topic of current major interest. Since the soot emitted is potentially carcinogenic, stringent legislation requiring reduction of its emission has been proposed in many countries and a post-treatment of the exhaust might therefore be required.

Conventional Methods. Soot emissions can be reduced by placing a filter in the exhaust pipe of an incinerator (as it was already proposed for diesel engines). To prevent pressure build-up, the accumulated soot particulates must be either removed from the filter, or burned *in situ* (or somewhere else). As the combustion temperature of the soot is much higher than the operational temperature found in the exhaust pipe, either the temperature in the exhaust pipe must be increased or the combustion temperature of the soot must be decreased by the addition of an oxidation catalyst in the form of fuel additives or by impregnation of the filter walls with an oxidation catalyst. However catalysts, still at the R&D phase, are easily poisoned and hard to use due to the large quantity of soot included in the exhaust gases.

An effort to develop an exhaust gas recycling system to reduce NO_x emission from diesel engines by lowering the combustion temperature in the engine cylinder failed 15 years ago. Although the primary goal was achieved, the recycling of soot-laden gases led to decreased engine power and increased the deposit on the valves, valve seats, piston surfaces and cylinder walls.

Plasma-Assisted Methods. Some thermal or non-thermal plasma processes were also proposed for carbonaceous materials processing :

It was shown in (*8*) that one can avoid soot formation in a very lean fuel combustion by the injection of a thermal plasma jet into the flame. This effect was realized by the addition of a small amount of electrical energy (about 4 % of the chemical throughput). It was also demonstrated that the formation of nitrogen oxide was reduced by using nitrogen gas as a plasma gas, and suggested that dissociated nitrogen atoms produced by plasma torch could work as reductants of nitrogen oxide. According to such a prediction, low concentration of nitrogen oxide could be established under the condition of low temperature and low oxygen gas content, otherwise formation of nitrogen oxide would prevail.

Cold oxygen plasma oxidation of coal was used for analytical purposes as a mild and selective procedure for organic matter determination (*9*). It was shown how excited gas molecules may be applied at 150°C instead of a non selective thermal oxidation with ordinary oxygen at 350-400°C. Anthracite (having a close to cyclic graphite structure) was easier to be oxidized than organics with linear carbon structures. However, the operation under vacuum (1-3 kPa), with a radio-frequency power supply (40.68 MHz), and at low power (no more than 40 W) make this plasma device quite unadapted to use for a larger gas flow cleaning from solid carbonaceous matter.

Ignition of solid coal is difficult to carry out (even though it is pulverized) without an auxiliary oil or gas burner. The use of a plasma jet (up to 18 kW, 100-300 A, 10 slm N_2 (standard liter per min) was recently studied (*10*) for the ignition of 150 g/min, -200 mesh pulverized coal combustion and inhibition of nitrogen oxide formation. The results show that :
- pulverized coal could be ignited with a plasma jet;
- the proportion of electrical energy is 6 to 8 % of the chemical throughput for a stable combustion of pulverized coal at ambient temperature;
- Nitrogen plasma can reduce the nitrogen oxide concentration from 800 ppm to 60 ppm while Hydrogen or Argon plasma does not.

A collaborative work (*11,12*) has resulted in the development of a quite complex experimental plasma reactor that can simultaneously remove NO_x and soot from diesel engine exhaust. A glass cylindrical plasma reactor of inner diameter 16.5 mm has the inner electrode made of a stainless metal screw of 6 mm diameter and 300 mm long, and the outer electrode is an aluminum foil wrapped around the glass tube. The reactor is energized by AC high voltage. The major innovation in this work is the addition of oil drops into the FG stream, which produces a dielectric mist and a more homogeneous discharge leading to the complete collection of soot (by an electrostatic precipitation to oil) as well as the removal of NO_x. To process part of a diesel engine exhaust the authors use 120 such tubular reactors in parallel which forms a battery of 450x312x560 mm size. The battery is powered by 24 loose wound transformers (15 kV, 20 mA, 50 Hz). The input power to 120 reactors at 12 kV was 1530 W (while installed power was 7200 VA). The energy given in the reactor to the exhausted gas is 76.5 kJ/m^3; the NO_x reduction efficiency is 0.75 kJ/m^3.ppm. The energy consumption for 100% soot collection is 1.5 GJ/kg.

GlidArc Principle and Applications

This simple and energy efficient non-thermal-plasma device (*13*) has been already success-fully applied in many processes studied since 1988. Its principle is schematically shown on Figure 1. At least two electrodes diverging with respect to each other are placed in a relatively fast gas flow (> 10 m/s) and in the flow direction. Gliding electric discharges are produced between the electrodes and across the flow. They start at the point where the distance between the electrodes is the shortest, and spread by gliding progressively along the electrodes in the direction of flow until they disappear after they have run a certain path. This path is defined by the geometry of the electrodes, by the conditions of flow and by the characteristics of the power supply (either AC or DC can be used (14,15)). Then, the electrical discharges immediately reform at the initial spot (16,17).

The fast displacement of the discharge roots on uncooled electrodes prevents their chemical corrosion or thermal erosion by usual high current arc establishment. The electrical energy is directly and totally transferred to the fast and turbulent gas flow. The average voltage ranges from 0.5 to 10 kV for currents from 0.1 to 3 A (per discharge). The instantaneous voltage, current and dissipated electric power data observed via digital oscilloscope (13) show almost random feature of the history of each gliding breakdown.

Transient electrical phenomena observed under near-to-atmospheric pressure are similar to the "corona discharge" type but at a much higher dissipated power level. A large ionized gas volume, obtained at low energy density, gives a non-equilibrium and reactive medium well adapted to run plasma-chemical reactions allowing efficient gas processing so that up to 45 % of the input electrical energy may be directly absorbed in an endothermic reaction. Physical model of the GlidArc shows that in such a plasma the exact notion of "temperature" cannot be used (18).

Almost any kind of gas or vapor, also heavily charged with solids or liquids, can be directly processed at no maintenance, at any inlet temperature, at 0.1-5 atm pressure range, and at negligible pressure drop : Ar, N_2, O_2, H_2, CO_2, CO, H_2S, SO_2, N_2O, NO, NO_2, CH_4, other HCs, chlorinated HCs, freons, different VOCs, ammonia, steam, air, and some of their mixtures like any FGs, either hot or cold.

GlidArc structure can either work as a classical burner or as a booster when processing very lean mixtures of following gases or vapors: H_2, CH_4, other HCs, VOCs as xylene, toluene, heptane, tetra-chloro-ethylene, methyl-ethyl-ketone (19-24), H_2S (20,22-28), ammonia, free or linked phenols, formaldehyde, and CH_3HS (22,25). GlidArc may therefore be applied :
- when an addition of classical fuel (requiring extra air with the nitrogen ballast) dilutes the fumes too much, increases greatly their volume and results in bigger absorption plants,
- to achieve higher flame temperature (29),
- to have a free choice of both temperature and red-ox conditions,
- when the gas mixture to be burnt has an insufficient content of combustible gases (which requires supply of extra energy),
- when the gas pulsation and/or velocity render the operation of the classical burner difficult and/or cause safety problems.

The GlidArc reactors can enhance many different processes, mostly in the engineering and/or environment control, like :
- conversion of the natural gas to the syngas H_2 + CO (30-32),
- Hydrogen and Sulfur recovery from concentrated H_2S or H_2S + CO_2 mixtures (SulfArc process), (20,23,26,27,33),
- FG purification from SO_x (34),

- methane transformation to acetylene and hydrogen,
- upgrading destruction of N2O (*35*),
- reforming of heavy petroleum residues,
- decomposition and acid recovery from concentrated freons,
- CO2 dissociation (*36*),
- overheating of steam (*37*), oxygen, and other gases,
- ignition of propellants,
- UV generation,
- decontamination of soil or industrial sands,
- activation of organic fibers,
- and other.

Current GlidArc reactors can be powered up to an industrial size, have two or more electrodes (3-phases, 6-phases, n-phases, parallel, serial or mixed mounting).

Combustion of Sooting Materials (*38*)

Two experimental installations have been used : a laboratory scale set-up for a preliminary check on a plasma assisted cleaning of the FG issued from nitro-benzene (NB) open burning and a pilot-scale installation devoted to do it with the TNT.

Nitro-Benzene Laboratory Experiment. The NB has been used at the laboratory in Orléans, France, for the obvious reason : this compound is non-explosive though it is quite close to the TNT as concerns its chemical composition and combustion property.

A vertical device similar to a petroleum lamp was used to a controlled combustion of the liquid NB. The lamp was put into a void supplied by a controlled flow of a cold or hot air. The NB combustion started using an electric igniter. The black fumes entered directly a two-stage GlidArc incinerator. This plasma incinerator was simply a 1/3 part of the six-stage GlidArc incinerator that we had to attach to the pilot-scale installation of the TNT combustion (see later).

The whole lamp mass difference (before and after an experiment), the air flow-rate, the inlet air temperature, temperatures at the entry and exit of GlidArc incinerator, and time were measured for each run. The covered ranges of these parameters (for a constant electric power) were the following ones :

total NB burnt :	6-33	g
time :	5-8.5	min
combustion rate (CR) :	0.9-4.4	g/min
air flow-rate (AFR) :	32-110	slm
temperature at the plasma reactor inlet (tI) :	70-280	°C
temperature at the plasma reactor exit (tE) :	85-282	°C

Only the total COVs contents in both treated and un-treated FGs were determined, after the soot removing (a solid filter), using a simplified gas-chromatograph with an empty, 1/8", 10 m metallic column, 40°C, and a total Flame Ionization Detector (FID) signal. A ratio of the signal from the non-treated FG (plasma reactor off) over the signal from the treated FG (plasma reactor on) gave us a Relative Cleaning Factor (RCF) for each run. Some results of the preliminary experiments are presented below :

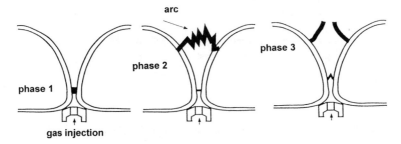

Figure 1. Start, life, and disappearance of the gliding arc.

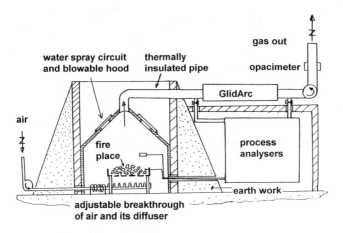

Figure 2. Waste treatment of explosive materials.

Run	4	5	6	7	8	10
CR (g/min)	1.3	1.2	1.1	0.9	4.4	3.7
AF (slm)	41	110	63	47	63	74
tI (°C)	190	170	118	145	280	70
tE (°C)	222	220	228	240	282	85
RCF	17	5	6	4	4	3

These results indicate that, for almost the same electric power of the GlidArc incinerator, one should work at the highest possible inlet FG temperature and the highest possible initial COV concentration in the gas to be treated. In such a way the plasma reactor will work as a simple continuous igniter of almost self-combustion of the organic pollutants.

Though the soot after-combustion was not measured during these experiments, a visual observation of the plasma incinerator action on the emitted FG clearly indicated a spectacular diminution of the fume darkness.

TNT Pilot Experiment. The pilot set-up for the combustion off gas processing (Figure 2) includes :
- elevated grate carrying the materials to be burnt,
- adjustable vaned orifice and its diffuser that is located under the grate,
- hood covering all the grate and having an exit of gas, liquid and/or solid particle effluent in its top,
- thermally insulated pipe connecting the hood top to the GlidArc burner, the last one being connected to a variable ventilator,
- measurement part : TV cameras and sensors connected to the installation and checking its parameters (thermometers, gas analyzers (spectrometer, NO_x, CO_2, CO, O_2 sensors, chemical analysis by absorption, gas chromatograph, UV Photoionization Organic Vapor Meter, and opacimeter),
- water spray circuit which may be operated by remote control in case of emergency; for the same reason the hatches fitting the hood can be ejected in case of accidental excessive pressure,
- room protected from possible explosions where operations of remote control are conducted; the same room can also be used for several destruction facilities.

Each section (of 6) of the GlidArc burner was composed of a 80 mm in. dia. quartz tube (250 mm long) in which 3 knife-shaped steel electrodes are put around the tube axis. The electrode gap is starting at 3 mm (ignition) to become about 70 mm at the electrode top (discharge disappearance). Six similar sections were assembled together, one after the other, so that the total length of this plasma incinerator was close to 2 m.

The whole reactor was connected to the high voltage, 50 Hz, 3-phase, multi-output power supply with current control (~ 1.3 A). The electric power injected to the reactor (1-9 kW) was carefully measured via both digital watt-meter and classical kWh-counter. The input gas flows (up to 200 m³/h) were also measured (via gas-velocity sensor) so the Specific Energy Input (SEI) could be precisely determined for each experiment.

Several experiments were performed for different TNT charge mass (a couple of kilogram range), FG initial composition and temperature (up to 300°C), gas flow-rate, and dissipated power. No chemical corrosion, no erosion and no short-circuits of GlidArc electrodes or their supports were observed during all experiments when processing the very smoky FG. A typical result related to the TNT combustion with plasma assisted FG cleaning is as follows :

FG flow rate :		50 m^3/h	
power dissipated in GlidArc burner :		3.2 kW	
FG composition	before burner	after burner	
CO	760 ppm	515 ppm	
NOx	1180 ppm	406 ppm	
NO2	206 ppm	158 ppm	
CO2	3.4 %	14.8 %	
O2	17.6 %	13.0 %	
soot	abundant	very little	

Conclusion

A high cleaning efficiency of the Flue-Gas (FG) from the sooting combustion of organic nitrates was achieved : almost complete disappearance of the soot (and the Products of Incomplete Combustion adsorbed on it) as well as an important lowering of the NOx and CO concentrations. This operation was performed at a relatively low energy cost of about 0.06 kWh per cubic meter of the treated FG.

New and very simple plasma-chemical process based on gliding electric discharge reactor can considerably reduce the nuisances of some combustion processes in which soot, Volatile Organic Compounds, Polycyclic Aromatic Hydrocarbon, NOx, CO, and other toxics are present in a FG. The use of such electrical method allows a full control of energy expense. High energy consuming and troublesome thermal or catalytic FG cleaning units can be substituted by this non-thermal and reactive plasma. Conditions of the GlidArc use are very flexible since it may be operated without any practical limit of pressure, temperature, flow-rate and initial composition of the FG. The process requires no particular FG pre-treatment and may be adapted to any size. It may be started, adjusted and stopped rapidly and easily. The pressure drop is low, and the reactor is very compact.

FG treatment is now operational when applied to the combustion of already tested explosive materials like TNT, HMX, RDX, TNT/RDX or TATB. In other cases, preliminary tests have to be conducted in order to adjust parameters. The unit could be scaled in size to treat varying quantities of wastes at various locations.

Hereby introduced new plasma-chemical process for the destruction of both soot and NOX may affect relevant rules and requirements that will be developed for waste combustion (and perhaps diesel exhausts) in the near future and may also add simplicity to related waste destruction facilities.

Literature Cited

1. Barfknecht, T.R. *Progr. Energy Combust. Sci.* **1983**, *8*, 199.
2. J.P. Longwell, J.P. In *Soot in Combustion Systems and its Toxic Properties*; Lahaye, J; Prado, G., Ed.; Plenum: New York, 1983, pp 37-56.

3. Lox, E.S.; Engler, B.H.; Koberstein, E. In *Catalysis and Automotive Pollution Control II;* Crucq, A., Ed.; Studies in Surface Science and Catalysis; Elsevier: Amsterdam, 1991, Vol. 71; p. 291.
4. Miller, J.A.; C.F. Melius, C.F. *Combustion and Flame* 1992, *91*, 21.
5. *Oil & Gas J.* 1993, Jan. 25, 29.
6. Frenklach, M. *Comb. Sci. and Tech.* 1990, *74*, 283.
7. Smedley, J.M.; Williams, A.; Bartle, K.D. *Combustion and Flame* 1992, *91*, 71.
8. Hilliard, J.C.; F.J. Weinberger, F.J. *Nature* 1976, *259*, 556.
9. Korobetskii, I.A.; Balabanova, N.V.;Popov, V.K.; Butakova, V.I.; Rus'janova, N.D. *Fuel* 1990, *69*, 654.
10. Takeda, K.; Hibi, M.; Takeuchi, H. *Proc. Jpn. Symp. Plasma Chem.*, 1992; Vol. 5; pp 297-302.
11. Fuji, K.; Higashi, M.; Suzuki, N. In *Non-Thermal Plasma Techniques for Pollution Control;* B.M. Penetrante, B.M.; S.E. Schultheis, S.E., Eds.; Electron Beam and Electrical Discharge Processing, NATO ASI Series G: Ecological Sciences, Springer-Verlag: Berlin, 1993, Vol. 34, Part B, pp 257-279.
12. Higashi, M.; Uchida, S.; Suzuki, N.; Fuji, K. In *Proc. 11th Int. Symp. on Plasma Chem.*, Harry, J.E., Ed.; Loughborough: England, 1993, Vol. 2; pp 704-709.
13. Chapelle, J.; Czernichowski, A. *Cahiers Français de l'Electricité, Industrie* 1992, *No. 4*, 20.
14. Harry, J.E.; Yuan, Q. In *Proc. 11th Int. Symp. on Plasma Chem.*, Harry, J.E., Ed.; Loughborough: England, 1993, Vol. 2, pp 734-739.
15. Czernichowski, A.; Janowski, J.; Stryczewska, H.D. In *Proc. 4th Int. Symp. on High Pressure Low Temperature Plasma Chem. (HAKONE IV)*, Bratislava, 1993; pp 111-116.
16. Lesueur, H.; Czernichowski, A.; Chapelle, J. *J. de Phys., Coll. de Phys., Suppl.* 1990, *18-C5*, 57.
17. Cormier, J.M.; Richard, F.; Chapelle, J.; Dudemaine, M. In *Proc. of 2nd Int. Conf. on Electr. Contacts, Arcs, Apparatus and their Applications*, Xi'an: China, 1993, pp 40-42.
18. Fridman, A.A.; Petrousov, A.; Chapelle, J.; Cormier, J.M.; Czernichowski, A.; Lesueur, H.; Stevefelt, J. *J. Phys. III France* 1994, *4*, 1449.
19. Czernichowski, A.; Lesueur, H.; Fillon, G. In *Proc. of Table Ronde "Procédés de Traitement des Déchets par Plasma"*, Odeillo: France, 1990, 2 pp.
20. Czernichowski, A.; Lesueur, H. In *Proc. of Plasma Applications to Waste Treatment, First Annual INEL Conference, Session II*, Idaho Falls: ID, 1991, 13 pp.
21. Czernichowski, A.; Lesueur, H. In *10th Int. Symp. on Plasma Chem.*, Bochum: Germany, 1991, paper 3.2.1, 6 pp.
22. Czernichowski, A.; Czech, T. In *Proc. of the IIIrd Int. Symp. on High Pressure, Low Temp. Plasma Chemistry (HAKONE III)*, Strasbourg: France, 1991, pp 147-152.
23. Czernichowski, A.. In *Coll. on "The Destruction of Waste and Toxic Materials Using Electrical Discharges"*, The Institution of Electrical Engineers, London, 1992, 5 pp.
24. Chapelle, J.; Czernichowski, A. *Rev. Gén. de l'Electr.* 1993, *No. 2*, 43.
25. Czernichowski, A.; Lesueur, H.; Czech, T.; Chapelle, J. In *10th Int. Symp. on Plasma Chemistry*, Bochum: Germany, 1991, paper 3.2.22, 6 pp.

26. Lesueur, H.; Czernichowski, A..; Granops, M. In *Emerging Technologies in Hazardous Waste Management V, ACS Special Symposium*, Tedder, D.W., Ed.; Atlanta: GA, 1993, Vol. I; pp 361-364.

27. Czernichowski, A. In *Non-Thermal Plasma Techniques for Pollution Control;* B.M. Penetrante, B.M.; S.E. Schultheis, S.E., Eds.; Electron Beam and Electrical Discharge Processing, NATO ASI Series G: Ecological Sciences, Springer-Verlag: Berlin, 1993, Vol. 34, Part B, pp 371-387.

28. Lesueur, H.; Czernichowski, A.; Granops, M. In *Proc. 11th Int. Symp. on Plasma Chem.*, Harry, J.E., Ed.; Loughborough: England, 1993, Vol. 2, pp 752-757.

29. Lesueur, H.; Czernichowski A.; Chapelle, J. In *Proc. IX Int. Conf. on Gaz Discharges and Their Applications*, Venice: Italy, 1988, pp 549-552.

30. Lesueur, H.; Czernichowski, A.; Chapelle, *J. de Phys., Coll. de Phys., Suppl.* **1990**, *18-C5*, 49.

31. Lesueur, H.;. Czernichowski, A.; Chapelle, J. *Int. J. Hydrogen Energy* **1994**, *19*, 139.

32. Meguernes, K.; Chapelle, J.; Czernichowski, A. In *Proc. 11th Int. Symp. on Plasma Chem.*, Harry, J.E., Ed.; Loughborough: England, 1993, Vol. 2, pp 710-715.

33. Sanijuk, S.V.; Kingsep, S.S.; Rusanov, V.D.; Czernichowski, A. In *Proc. 11th Int. Symp. on Plasma Chem.*, Harry, J.E., Ed.; Loughborough: England, 1993, Vol. 2, pp 740-745.

34. Czernichowski, A.; Polaczek, J.; Czech, T. In *Proc. 11th Int. Symp. on Plasma Chem.*, Harry, J.E., Ed.; Loughborough: England, 1993, Vol. 2, pp 674-679.

35. Czernichowski, A.; Gorius, A.; Charamel, A.; Potapkin, B.V. In *Proc. of The First Int. Conf. on Advanced Oxidation Technologies for Water and Air Remediation*, London: Ontario, 1994, pp 247-248.

36. Czernichowski, A.; Ranaivosoloarimanana, A.; Janowski, T.; Stryczewska, H.D.; Cojan, M.; Fridman, A.A.; In *Proc. of Int. Conf. on Electromagnetic Devices and Processes in Environment Protection ELMECO'94*, Kazimierz Dolny: Poland, 1994, p. 71.

37. Czernichowski, P.; Czernichowski, A. In *9ème Colloque Université-Industrie "Les techniques électriques et la qualité du séchage"*, Bordeaux-Talence: France, 1994, pp. B1-1 - B1-7.

38. Labbé, Ph.; Czernichowski, A.; Laval, F.; Foucher, Ph.; Lesueur, H.; Chapelle, J. In *Emerging Technologies in Hazardous Waste Management V, ACS Special Symposium*, Tedder, D.W., Ed.; Atlanta: GA, 1993, Vol. II; pp 399-402.

RECEIVED March 14, 1995

CHEMICAL OXIDATION AND CATALYSIS

Chapter 13

Air–Nitric Acid Destructive Oxidation of Organic Wastes

James R. Smith

Savannah River Technology Center,
Westinghouse Savannah River Company, Aiken, SC 29802

Many organic materials have been completely oxidized to CO_2, CO, and inorganic acids in a 0.1M HNO_3/14.8M H_3PO_4 solution with air sparging. Addition of 0.001M Pd^{+2} reduces the CO to near 1% of the released carbon gases. To accomplish complete oxidation the solution temperature must be maintained above 130-150°C. Organic materials quantitatively destroyed include neoprene, cellulose, EDTA, TBP, tartaric acid, and nitromethane. The oxidation is usually complete in a few hours for soluble organic materials. The oxidation rate for non-aliphatic organic solids is moderately fast and surface area dependent. The rate for aliphatic organic compounds (polyethylene, PVC, and n-dodecane) is relatively very slow. This is due to the large energy required to abstract a hydrogen atom from these compounds, 99 kcal/mole. The combination of $NO_2\cdot$ and $H\cdot$ to produce HNO_2 releases only 88 kcal/mole. Under conditions of high $NO_2\cdot$ concentration it should be possible to oxidize these aliphatic compounds.

The development of a general process for the destructive oxidation of pure organic compounds could have many applications for environmental cleanup. A liquid phase oxidation process should not produce any ash making the system more environmentally contained. A simple process that uses oxygen from air, or another readily available cheap oxidant as the net oxidant, would be relatively inexpensive per unit of waste consumed. This work represents studies into development of such a process. Nitric acid is used as a catalyst and oxidant since it can be regenerated by air in an acid recovery system and to some extent in the reaction solution.

Liquid phase oxidation of organic molecules should be easier than gas phase oxidation at a given temperature (1). This is due to the ability to produce high concentrations of the reactants and to the lessening of termination since radicals have a harder time diffusing to the walls. Direct oxidation of most organic compounds by HNO_3, nitric acid, is energetically favorable but very slow due to its inability to break the carbon-hydrogen bond (1,2). The following heats of reaction, ΔH, values (in kcal/mole) are calculated, or approximated, using Table I.

0097–6156/95/0607–0156$12.00/0

$$RCH_3 + HNO_3 \longrightarrow RCH_2OH + HNO_2 \qquad \Delta H \cong -25$$

$$RCH_3 + HNO_3 \longrightarrow RCH_2\bullet + H_2O + NO_2\bullet \qquad \Delta H^{\cdot} \cong 35$$

$$RCH_3 + HNO_2 \longrightarrow RCH_2\bullet + H_2O + NO\bullet \qquad \Delta H^{\cdot} \cong 28$$

Table I. Heats of Formation for Molecules and Radicals at 25°C (1,2)

Compound	$\Delta H^{\cdot}_{form}$ (kcal/mole)	Compound	$\Delta H^{\cdot}_{form}$ (kcal/mole)
CH_3CH_3	-20	[a]CH_3NO_2(aq)	-8
CH_3CH_2OH(aq)	-66	HCO_2H(aq)	-98
CH_3OH(aq)	-59	$NO_2\bullet$	8
CH_3CHO(aq)	-50	HNO_2(aq)	-28
$CH_3CH_2\bullet$	26	HNO_3(aq)	-49
[a]$CH_3(O)C\bullet$	-21	$NO\bullet$	22
[a]CH_3CH_2OOH	-35	H_2O	-68
[a]$CH_3CH_2OO\bullet$	23	CO_2(aq)	-98
$CH_3\bullet$	31	NH_2OH (aq)	-22

[a] Approximated using other values (1,2)

R denotes an organic group unaffecting the $\Delta H^{\cdot}$ for the shown reaction. The oxidation of organic compounds is usually initiated by the production of organic radicals generated by dissolved $NO_2\bullet$ and $NO\bullet$ in solution. For many types of organic compounds the attack by $NO_2\bullet$ can be first order.

$$CH_3(OH)CH_2 + NO_2\bullet \longrightarrow CH_3(OH)CH\bullet + HNO_2 \qquad \Delta H^{\cdot} \cong 0$$

$$CH_3CHO + NO_2\bullet \longrightarrow CH_3(O)C\bullet + HNO_2 \qquad \Delta H^{\cdot} \cong -7$$

For aliphatic compounds high concentrations of $NO_2\bullet$ and $NO\bullet$ are needed.

$$RCH_3 + H_2O + 3NO_2\bullet \longrightarrow RCH_2\bullet + 2HNO_2 + HNO_3 \qquad \Delta H^{\cdot} \cong -15 \qquad (1)$$

$$RCH_3 + H_2O + 2NO_2\bullet + NO\bullet \longrightarrow RCH_2\bullet + 3HNO_2 \qquad \Delta H^{\cdot} \cong -8 \qquad (2)$$

A typical aliphatic carbon-hydrogen bond strength of 99 kcal/mole was used in the calculations (2). Adding oxygen to the solution, by air sparging, can set up a radical propagating oxidation mechanism. Organic radicals quickly react with molecular oxygen to form a peroxic radical.

$$RCH_2\bullet + O_2 \longrightarrow RCH_2OO\bullet \qquad \Delta H^{\cdot} \cong -3 \qquad (3)$$

The peroxic radical can easily abstract a hydrogen from even an aliphatic organic molecule since an oxygen-hydrogen bond is formed to break a carbon-hydrogen bond. This results in an organic hydroperoxide and another organic radical. Above 130-150°C the organic hydroperoxides decompose to release H_2O or CO_2 (1).

$$RCH_2OO\bullet + RCH_3 ---> RCH_2OOH + RCH_2\bullet \qquad \Delta H\dot{=}-12 \qquad (4)$$

$$RCH_2OOH ---> RCHO + H_2O \qquad \Delta H\dot{=}-83 \qquad (5)$$

The organic radicals can also be oxidized by nitric and nitrous acids or nitrated by $NO_2\bullet$.

$$RCH_2\bullet + HNO_3 ---> RCH_2OH + NO_2\bullet \qquad \Delta H\dot{=}-35 \qquad (6)$$

$$RCH_2\bullet + HNO_2 ---> RCH_2OH + NO\bullet \qquad \Delta H\dot{=}-42 \qquad (7)$$

$$RCH_2\bullet + NO_2\bullet ---> RCH_2NO_2 \qquad \Delta H\dot{=}-52 \qquad (8)$$

Oxidation of the carbon-carbon bond is also possible but slow probably due to steric factors.

$$CH_3CH_3 + NO_2\bullet + H_2O ---> 2CH_3OH + NO\bullet \qquad \Delta H\dot{=}-16$$

Methods and Materials

Weight measurements were taken using a Sartorious (Handy) balance. All chemicals were of at least Reagent grade quality meeting ACS specifications except for the cellulose (Whatman 40 ashless filter paper) and neoprene (glovebox glove made by Siebe North, Inc., Charleston, SC). Gas samples were analyzed on a Varian 3400 GC using Molecular Sieve-13X and Chromosorb 106 columns.

The data collected was generated from two similar but different oxidation systems. The first used a peristaltic pump to circulate air through a closed system starting with (and continuing in order) a three-liter reaction vessel containing a liter of reactant solution, an ice trap, a 26.6 liter polypropylene bottle, a 30 ml gas sample vessel, and then re-entering the peristaltic pump. The system was connected to an inverted burette, in acidic solution, to measure changes in the gas volume of the system. The total gas volume of the system was 29.0 ± 0.2 liters. The air entered the reaction solution through a three-inch diameter medium glass frit plate generating very small bubbles in the solution. The air circulation rate for this system was maintained at 500 ml/minute. The all-glass three-liter reaction vessel was set in a heating mantle and had four ground glass openings which provided access for a thermometer, an air cooled condenser (air outlet), air sparge inlet, and sample introduction-holder port. Gas samples were taken at the beginning, end, and during the oxidation reaction. Carbon balance results gave a combined error of $\pm3\%$.

The second system was of a flow-through design. A peristaltic pump was used to push air through the reaction system. The incoming air was stripped of CO_2 using Ascarite-II (Thomas Scientific). The air was delivered to the reaction vessel at 100 ml/minute. The reaction vessel was the same as described above except that a smaller glass frit sparger was used. The gas stream leaving the reaction vessel was scrubbed of $NO_2\bullet$ by a 0.5M sulfamic acid solution. The products of the $NO_2\bullet$ reaction with the sulfamic acid were nitric acid (stays in solution) and N_2 gas. The gas stream then entered approximately 100 grams of Ascarite-II in a polypropylene bottle. Weighing the bottle before and after the reaction determined the weight of CO_2 produced from the oxidation. The gas stream then entered a column of Pd metal on a Kieselguhr support maintained above $140°C$. The Pd/Kieselguhr catalyzed the air oxidation of any CO in the gas stream to CO_2. The generated CO_2 is then absorbed and weighed on a second bottle of Ascarite-II. Calibration of this system for the absorption of CO and CO_2 was accomplished by adding a weighed amount of dried sodium oxalate to concentrated sulfuric or phosphoric acid. In the absence of oxidizing compounds an equal molar amount of CO and CO_2 was formed. Carbon balances from the calibration determined a combined accuracy of ±2%.

Results and Discussion

The onset of oxidation for the soluble organic compounds was about $120^\circ C$. Gram quantities of organic material was oxidized per run. At $140^\circ C$ the oxidation was complete in less than fifteen minutes ($NO_2\bullet$ stopped being released from the solution). The results of the carbon balance studies are tabulated in Table II. Complete oxidation, within experimental error, was obtained for the compounds listed (with the exception of tartaric acid) for at least one of the runs. Tartaric acid should also be completely oxidizable. The purpose for the oxidation of tartaric acid was to determine the fraction of CO released. The results show that a more highly oxidized compound, such as tartaric acid, releases a smaller fraction of CO. The production of CO was very pronounced for TBP (tributylphosphate) and nitromethane. The result for nitromethane was not surprising since it should be quickly hydrolysed to formic acid in a strong mineral acid at these temperatures (3).

$$CH_3NO_2 + H_2O + H_3PO_4 \text{ ---> } HCO_2H + H_2NOH\bullet H_3PO_4 \qquad \Delta H^\cdot \cong -44$$

$$H_2NOH\bullet H_3PO_4 + HNO_2 \text{ ---> } N_2O + 2H_2O + H_3PO_4 \qquad \Delta H^\cdot \cong -66$$

Dehydration of formic acid to CO and water is slightly endothermic (4 kcal/mole) (2) but probably quick in 14.8M H_3PO_4 which is a strong dehydrating agent. The strong dehydrating ability of the reaction solution is probably aiding in the decomposition of the organic oxidation products. Cellulose is rapidly carbonized at around $140^\circ C$ to form carbon and water.

$$RCH(OH)CO_2H \text{ ---> } H_2O + CO + RCHO$$

$$C_6H_{10}O_5 \text{ (cellulose) ---> } 6C + 5H_2O$$

The carbon formed is easily attacked by the nitric acid. It is possible that the relative production of CO and CO_2 is determined by competing mechanisms; the CO by a dehydration mechanism and the CO_2 produced by oxidation with HNO_3 and $NO_2\bullet$.

$$CH_3CHO + NO_2\bullet \text{ ---> } CH_3(O)C\bullet + HNO_2 \qquad \Delta H^\cdot \cong -7$$

$$CH_3(O)C\bullet + HNO_3 \text{ ---> } CH_3\bullet + CO_2 + HNO_2 \qquad \Delta H^\cdot \cong 4$$

The slightly endothermic $\Delta H^\cdot$ for this last reaction shows how dehydration to produce CO can compete with oxidation to produce CO_2.

Carbon-nitrogen bonds are relatively weak (~75 kcal/mole) (1) so complete oxidation of EDTA (ethylene-diamine-tetraacetic acid) was not surprising. Oxidation of TBP was performed to test the ability to destroy nearly aliphatic compounds. Butanol, a hydrolysis product of TBP, resembles an aliphatic compound except for the alcohol group. The weakening of the adjacent carbon-hydrogen bonds by this lone oxygen seems to be sufficient to start the chain oxidation process. Oxidation of compounds of this type have been said to act like a candle that has been lit on one end (1). The carbon atoms are oxidized in order down the chain.

The surface oxidation of neoprene (poly [2-chloro-1,3-butadiene]) was found to be even, allowing measurement of the surface area and weight loss during its destruction. The oxidation rate at varying temperatures has been measured for this compound, Figure 1. The activation energy, E_a, was determined to be 22.9 kcal/mole

Table II.   Carbon Balance for Oxidation of Various Organic Compounds in a 0.05-0.1M HNO_3/14.8M H_3PO_4 Solution, 120-160°C

Compound	Co-catalyst Metal (conc.)	Percentage Carbon Released as CO_2 and CO	Percentage CO of CO_2 and CO
cellulose	none	98±2[a]	20±1
"	Pd (0.0012M)	99±2	0.9±1
"	Rh (0.003M)	97±3[a]	6.3±1.5
EDTA	none	99±2	25±1
"	Pd (0.0012M)	>91±2	0.9±1
"	Rh (0.003M)	>84±3[a]	19±1.5
TBP	none	103±3[a]	43±1[b]
"	Pd (0.0012M)	-	1.3±1
nitromethane	none	101±2	60±1
"	Pd (0.0012M)	-	2.3±1
tartaric acid	none	>96±2	15±1
neoprene	Pd (0.001)	101±10	<17

[a] GC analysis observed no H_2, methane, or ethane.

[b] Determined from butanol (a hydrolysis product of TBP)

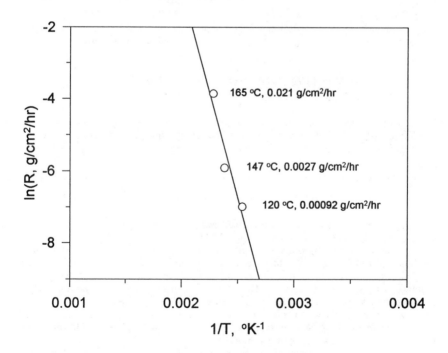

Figure 1.   Oxidation Rate, R, for Neoprene in 0.1M HNO_3/14.8M H_3PO_4 Solution at Varying Temperature

with a frequency factor, A, of $7.68E11$ s^{-1}. The release of chloride during the oxidation had no observable effect at the low concentration, up to 0.1M, formed by the destruction of the neoprene. Oxidation of neoprene is relatively rapid due to carbon-carbon double bond weakening of carbon-hydrogen bonds in a positions relative to the double bond (*1*).

The oxidation of aliphatic compounds such as polyethylene, PVC (polyvinylchloride), and n-dodecane was found to be unmeasureably slow in this reaction solution below $180°C$. In an air sparged system the concentration of $NO\cdot$ and $NO_2\cdot$ should be too low for reactions (1) and (2) to have any appreciable effect. In near 0.1M HNO_3 solution reactions (3), (4), and (5) would be hampered by destruction of the organic radicals by reactions (6), (7), and (8). It should be possible though under the right reaction conditions to oxidize aliphatic compounds by reactions (3), (4), and (5) or (6), (7), and (8) in this temperature range (*1*).

Two platinum group metals were studied as co-catalysts, palladium and rhodium. Palladium performed better in reducing the percentage of organic carbon released from solution as CO. The palladium metal formed during the oxidation of CO is reoxidized back to Pd^{+2} by nitric acid.

$$CO + Pd^{+2} + H_2O ---> CO_2 + Pd + 2H^+$$
$$Pd + 2HNO_3 --->Pd^{+2} + H_2O + 2NO_2^-$$

Unlike palladium, rhodium formed a significant amount of N_2O when used as a co-catalyst. Rhodium catalyzes the $NO\cdot$ oxidation of CO to produce CO_2 and N_2O^4. Even though the extent to which palladium is aiding in oxidation of the organic materials in solution is unknown, palladium is known to be an effective catalyst for oxidation of many types of organic compounds (*5*).

During some runs when an air sparge rate of 500 ml/minute was used, the solution temperature had to be raised to near $150°C$ to complete the oxidation. This is probably due to the formation of hydroperoxide compounds which decompose in the $130\text{-}150°C$ range (*1*). The extent to which the oxygen is interacting with the organic radicals is unknown. Oxygen could also enter as an oxidant by oxidizing dissolved $NO\cdot$ to $NO_2\cdot$. Even though the self-decomposition of organic hydroperoxides have a high activation energy (*1*), the concentration of these compounds are self limiting.

$$RCH_2OOH + R'CH_2OOH ---> RCHO + R'CHO + 2H_2O \qquad \Delta H\overset{.}{\cong}-166$$

Conclusions

Complete oxidation of many types of organic compounds is possible in a 0.1M $HNO_3/14.8M$ H_3PO_4 solution with air sparging. Oxidation of aliphatic compounds will require different and maybe more extreme conditions (i.e, pressurization and high nitric acid concentrations). Trace amounts of palladium should be capable of minimizing the production of CO to near 1% of the released carbon gases. Use of palladium should also minimize the production of N_2O which can not easily be reoxidized to HNO_3.

The released HNO_3, NO, and NO_2 can be reoxidized in a standard air driven acid recovery system providing a cheap source of oxidant. There should be no ash produced from the process making it more environmentally contained. There are existing resins that can be used to absorb hazardous metals such as mercury, silver, lead, and cadmium that work well in phosphate solutions (*6*). The reaction conditions were relatively mild, ambient pressure and $<180°C$, making the process potentially easy and cheap to implement.   Phosporic acid, H_3PO_4, and its solutions only mildly attack common stainless-steels (such as 304-L), especially at room temperature[7]. This would make handling and containing process solutions in plumbing and tanks safe and inexpensive.

Notation

$\Delta H^{\cdot}$ - standard heat (enthalpy) of reaction, $\Delta H^{\cdot \circ}{}_{298}$
R - an organic group unaffecting the $\Delta H^{\cdot}$ for the shown reaction.
neoprene - poly(2-chloro-1,3-butadiene)
EDTA - ethylenediaminetetraacetic acid
TBP - tributylphosphate
tartaric acid - 2,3-dihydroxybutanedioic acid
Acsarite-II - solid NaOH on a solid support (for CO_2 absorption).
Kieselguhr - diatomaceous earth (silica from the skeletons of diatoms).

Acknowledgements

This work was supported by the Department of Energy under Contract No. DE-AC09-89SR18035.   I would like to acknowledge David T. Hobbs of SRTC/WSRC for his gas sample analysis support.

Literature Cited

1.   Seminov, N.N. *Some Problems in Chemical Kinetics and Reactivity;* Princeton Univ. Press: Princeton, NJ, 1958; Vol. 1 and 2.
2.   Dickerson, R.E. *Molecular Thermodynamics;* The Benjamin/Cummings Publ. Co.: Menlo Park, CA, 1969.
3.   Fuson, R.C. *Advanced Organic Chemistry*; John Wiley and Sons, Inc.: New York, NY, 1950, p. 507.
4.   Dickson, R.S. *Organometallic Chemistry of Rhodium and Iridium*; Academic Press, Inc.: New York, NY, 1983, p. 7.
5.   Maitlis, P.M. *The Organic Chemistry of Palladium*; Academic Press, Inc.: New York, NY,1971.
6   Osreen, A.B.; and Bibler, J.P. *Water, Air, Soil Polluti.* **1991,** *58,* 63-74.
7.   *Corrosion Resistance Tables;* Schweitzer, P.A., Ed.; 3rd Edition; Marcel Dekker, Inc.: New York, NY, 1991.

RECEIVED June 6, 1995

Chapter 14

Oxidative and Catalytic Removal of Hydrogen Sulfide from Spent Caustic Liquors by Manganese Compounds

Bhupendra R. Patel and Philip A. Vella

Carus Chemical Company, 1001 Boyce Memorial Drive, P.O. Box 1500, Ottawa, IL 61350

In the presence of air, the oxidation of sulfide by potassium permanganate occurs by two paths. The first is the direct oxidation of sulfide by permanganate. From this reaction, manganese dioxide is formed. In the presence of air (oxygen) the MnO_2 acts as an oxygen transfer agent resulting in further sulfide oxidation. This effect was proven by experiments carried out in a nitrogen atmosphere with MnO_2 as the sole potential oxidant. Through these pathways, less than the theoretical amount of $KMnO_4$ is needed for complete sulfide oxidation. The above trends were observed in both clean and spent caustic matrices.

Spent caustic liquors containing high sulfide concentrations are generated in the petrochemical and petroleum refinery industries. In particular, the petroleum industry uses caustic scrubbers to remove acidic gases during the manufacture of ethylene. In 1992, more than 40 billion pounds of ethylene were produced in the United States (*1*). The manufacturing process involves thermal cracking of ethane, propane, butane, or naphtha. Severe reaction conditions (900°C) cause the spontaneous homolysis of C-C and C-H bonds. After cracking, the resulting gases are contacted with a caustic solution to remove unwanted compounds such as hydrogen sulfide and carbon monoxide which are present in very high concentrations. A poor quality feed material is sometimes a source of hydrogen sulfide. In addition, some ethylene manufacturers add reduced sulfur compounds (to create a reducing environment) which contribute to the formation of H_2S. These gases are retained as sodium sulfide and sodium carbonate in the spent liquor. Caustic scrubbing also removes mercaptans, phenols and other organic compounds which dissolve or become emulsified in the caustic solution.

Because of the presence of reactive sulfide, spent caustic liquor is classified as a hazardous waste under the Resource Conservation and Recovery Act (RCRA) (*2*). In the past, deep well injection was an accepted disposal technology. However, due to increasingly stringent RCRA regulations (the material must be "deactivated"), this method is no longer favorable. Alternatives to deep well injection include wet air oxidation, chemical oxidation, and dilution followed by biological oxidation. Of these, complete chemical oxidation is usually an expensive alternative.

0097–6156/95/0607–0163$12.00/0

Potassium permanganate is often used for sulfide oxidation in municipal waste treatment plants and collection systems. The relatively short reaction time of potassium permanganate is a major advantage in treating hydrogen sulfide. Another advantage of potassium permanganate treatment is the formation of the catalytic compound manganese dioxide. According to Stewart (3), permanganate reacts with sulfide according to Equation 1.

$$8KMnO_4 + 3H_2S \text{ ------->} 3K_2SO_4 + 2KOH + 2H_2O + 8MnO_2 \qquad (1)$$

At an alkaline pH, the permanganate is reduced to manganese dioxide and the sulfide is oxidized to sulfate. Any remaining sulfide can then react with the MnO_2 to form elemental sulfur and MnO (Equation 2). In the presence of an oxygen source, MnO_2 can be regenerated (Equation 3) and would be available for additional sulfide oxidation. With the initiation of reaction 1, reactions 2 and 3 proceed simultaneously. As a result of these three reactions, less than the theoretical amount of $KMnO_4$ should be needed for sulfide oxidation.

$$2MnO_2 + S^{2-} \text{ ------->} 2MnO + S^0 + O_2 \qquad (2)$$

$$MnO + 1/2O_2 \text{ ------->} MnO_2 \qquad (3)$$

Studies performed under anoxic conditions showed that only partial sulfide removal occurs with the addition of less than the stoichiometric amount of permanganate (4). However, field observations of sulfide removal in municipal wastewater systems offer a different perspective. They support the concept of sub-stoichiometric amounts of permanganate for efficient sulfide oxidation. These municipal conditions differ from the laboratory study with respect to the fact that oxygen is available (air diffusion during normal flow and mixing). It is proposed that the large surface area of the MnO_2 (due to hydration) is responsible for its oxygen transfer characteristics (5) and subsequent oxidation of sulfide.

This paper will compare the oxidative capacity of $KMnO_4$ for the removal of sulfide. The study will encompass both a clean and a high strength sulfide spent caustic system. Special attention will be directed to the effects of air (oxygen) on these reactions. It will be shown that MnO_2, formed from the initial oxidation of sulfide, can catalytically oxidize sulfide adding to the efficiency of permanganate. By using $KMnO_4$ under the proper conditions, a more cost effective route to sulfide oxidation, compared to direct oxidation with permanganate, can be achieved.

Experimental.

Reagents and Chemicals. Sulfide standards were prepared from sodium sulfide nine hydrate ($Na_2S \cdot 9 H_2O$) obtained from EM Science and maintained at pH 10 using an ammonium chloride/ammonium hydroxide buffer. Reagents for the methylene blue method were obtained from Hach Co. Stock solutions of potassium permanganate were prepared from free flow grade CAIROX potassium permanganate. Potassium permanganate standards were prepared from EM Science Dilut-It analytical reagent concentrate and used to generate a calibration curve. Grade 5 nitrogen was obtained from Airco Gas Co. Manganese dioxide was freshly precipitated according to a procedure using reagent grade manganous sulfate and free flow grade potassium permanganate (6). The resulting mixture was found to be free of residual permanganate. The filtrate was analyzed for residual manganese by atomic absorbance spectroscopy and showed 5 mg/L or less dissolved manganese.

Methods of Analysis and Data Collection. The sulfide stock solution was analyzed using the iodine thiosulfate titration method 4500-S-2 E as described in Standard Methods (7). The methylene blue method 4500-S-2 (sulfide determined spectrophotometrically) was used as a primary calibration method for dilute sulfide analysis and used to verify the data generated from an Ion Selective Electrode (ISE). From this data, a calibration curve was made by plotting $\log[S^{2-}]$ vs mV obtained by the ISE. An equation was then developed which allowed the calculation of sulfide concentrations based on millivolt values. This analytical technique is suitable for the continuous monitoring of sulfide concentrations (8). Potassium permanganate concentrations were obtained on a Shimadzu UV-160 spectrophotometer at 525.3 nm.

An Orion 9416 Ion Selective Electrode (Ag/Ag_2S) in conjunction with Orion 9002 double junction reference electrode ($AgCl/KNO_3$) was used to measure mV in the sulfide bearing solutions. Electrodes were cleaned before each run according to the manufacturer's instruction to eliminate any electrode coating. The millivolt signals obtained by the ISE were recorded on an OM-160 data logger from Omega Engineering Inc. using a model 13b data processing module. The recording intervals ranged from 5 to 30 seconds for each experiment. The recorded data were transferred into an IBM compatible computer with provided software to create the database files. Further data processing was performed on a Macintosh computer system.

General Test Conditions. Batch experiments were performed in a well-stirred round bottom 500 mL 3 neck flask. This configuration allowed for placement of the electrodes and gas diffusion tube. The initial sulfide concentration was approximately 10 mg/L in the low sulfide study and approximately 500 mg/L in the high strength case. The pH of these solutions was maintained at 10 during the experiment. The starting and ending concentrations were determined by the ISE and confirmed by the methylene blue method. To evaluate the effects of air on sulfide oxidation, either air or nitrogen was bubbled through the test sample. The flow of either gas was controlled by a mass flow controller at either 93 mL/min (low level sulfide) or 600 mL/min (high level sulfide). The pH of the spent caustic solution was 12-13. The air flow in these experiments was maintained at 600 mL/min.

Results and Discussion.

Sulfide Oxidation with Potassium Permanganate: Clean Water Matrix.

Low Level Sulfide. Based on Equation 1, the theoretical weight ratio of permanganate to sulfide is 12.4:1 for complete oxidation under alkaline conditions. To determine if this ratio is valid, sulfide solutions were treated with potassium permanganate in the presence of air. The weight ratios of $KMnO_4$ to S^{2-} examined were 1:1, 2:1, and 4:1. The results showed that at a 4:1 weight ratio, 99.9% of the sulfide was removed in less than 1 minute (Figure 1). This is significantly lower than the amount expected based on Equation 1.

This result indicated that some factor other than the direct oxidation by $KMnO_4$ was aiding in sulfide removal. To evaluate the possible effects of air on sulfide oxidation with potassium permanganate, a set of experiments was performed in a nitrogen environment. The sulfide solution was purged with nitrogen for 15 minutes to drive off any oxygen present in the solution or reaction flask. A continuous flow of nitrogen was then maintained throughout the experiment. It should be noted that the effects of stripping with nitrogen were investigated and no

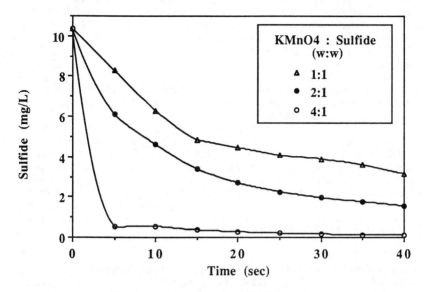

Figure 1. Sulfide Removal with Potassium Permanganate and Air

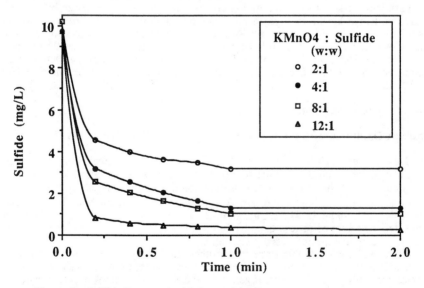

Figure 2. Sulfide Removal with Potassium Permanganate and Nitrogen

substantial loss was observed. The effect of air alone resulted in less than a 10% loss of sulfide concentration over a 10 minute period.

Figure 2 shows sulfide removal with $KMnO_4$ in the absence of air. The dosages used were 2:1, 4:1, 8:1, and 12:1 potassium permanganate to sulfide (w:w). All reaction times were kept at 10 minutes to ensure a complete reaction. (Note: Only 2 minutes are shown. No further sulfide oxidation was observed after this time.) After the first minute, no residual permanganate was left in any of the reactions. In the absence of air, higher dosages of permanganate were required. To remove 99.9% of the sulfide, more than a 12:1 permanganate to sulfide weight ratio was required. These findings are in the agreement with the work reported by Cadena and Peters (9) and follow the stoichiometry given in Equation 1.

Treatment with Freshly Precipitated Manganese Dioxide. It has been reported that MnO_2 has been used to treat tannery waste for sulfide odors (10). In order to confirm the effect of MnO_2 with air and further substantiate its oxidative capability, sulfide solutions were treated with freshly precipitated MnO_2. These results are presented in Figure 3. Although sulfide oxidation does occur, it is at a much slower rate compared to the reaction with potassium permanganate. However, total oxidation can be achieved either by increasing the MnO_2 dose or increasing the reaction time. When this reaction was carried out in a nitrogen environment, no significant oxidation was observed (Figure 4). The limiting factors for MnO_2 treatment are the longer reaction times and the requirement of oxygen (air) with agitation to insure effective oxygen transfer.

From Figures 1-4, the interaction of manganese dioxide, air, and sulfide has been demonstrated. The data support the idea that sulfide oxidation with potassium permanganate is a two step reaction. The first step is the direct oxidation of sulfide by permanganate forming MnO_2 (Equation 1). This freshly formed MnO_2 has a hydrous structure and acts as an oxygen transfer agent in alkaline conditions. The manganese dioxide reacts with sulfide and is reduced to manganous oxide (Equation 2). Manganese (II) oxide reacts with atmospheric or dissolved oxygen to reform manganese dioxide (Equation 3). The reformed MnO_2 can then continue to react with any remaining sulfide present in solution. These steps occur simultaneously during the sulfide oxidation with air present resulting in an overall increase in the apparent oxidative capacity of $KMnO_4$.

Oxidation of High Strength Sulfide. The next step in the investigation was to determine if the effects observed at low sulfide concentrations were similar at high sulfide concentrations. Sulfide concentrations of 500 mg/L were prepared in a clean matrix and the effects of air and $KMnO_4$ dose were studied. The results of those studies are presented in Figures 5 and 6. As seen from Figure 5, extremely low weight ratios of $KMnO_4$ to sulfide (0.16:1) are able to remove over 90% of the sulfide within 40 minutes. This is considerably lower than that observed at low sulfide concentrations (4:1) and well below the theoretical amount of 12:1. The effect of air flow was also determined. Based on the data shown in Figure 6, there does not appear to be any significant difference in a 10 fold increase in air if the original $KMnO_4$ concentration is sufficient. This is evident by comparing the sulfide removal at a $KMnO_4$ dose of 0.16:1 with 60 or 600 mL/min of air flow. However, at low $KMnO_4$ concentrations, a significant effect of air on sulfide removal is noticed. This indicates that at these levels, oxidation of sulfide by MnO_2 is predominant.

Sulfide Oxidation with Potassium Permanganate: Spent Caustic Waste. Having shown the effects and interactions of $KMnO_4$, MnO_2 and air on low and high sulfide concentrations in a clean matrix, an investigation into their use for the

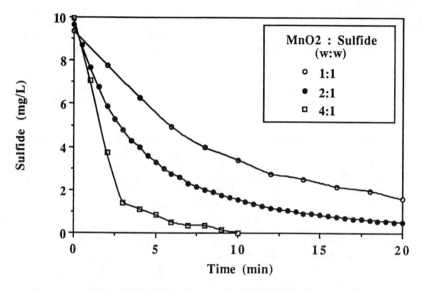

Figure 3. Sulfide Removal with MnO_2 and Air

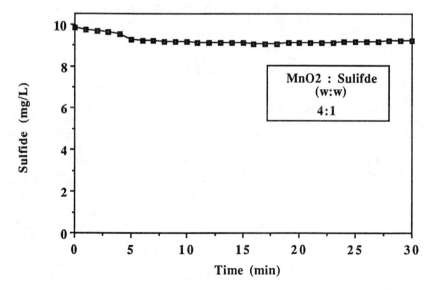

Figure 4. Sulfide Removal with MnO_2 and Nitrogen

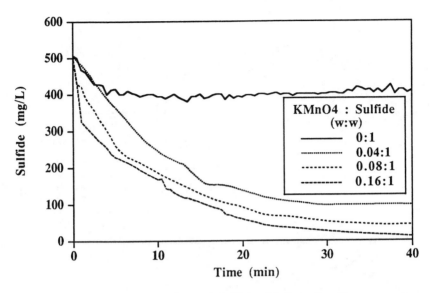

Figure 5. Oxidation of High Strength Sulfide with Potassium Permanganate

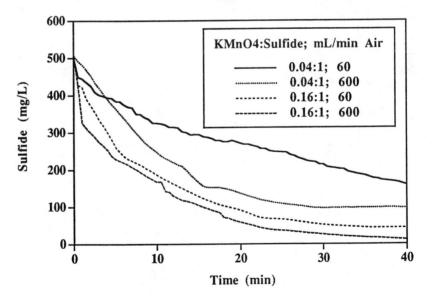

Figure 6. Effects of Air Flow and KMnO4 Dose on Sulfide Removal

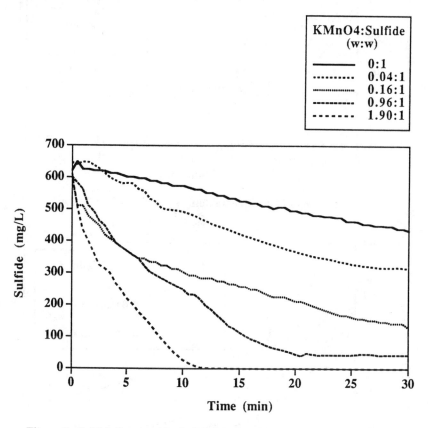

Figure 7. Sulfide Removal in Spent Caustic with Potassium Permanganate

oxidation of a high strength sulfide waste was initiated. A waste sample obtained from an ethylene producer had a pH of 12.8 and an initial sulfide concentration of 625 mg/L. Oxidation was carried out with varying amounts of $KMnO_4$ at an air flow rate of 600 mL/min. The results of this experiment are given in Figure 7. Under these conditions, a 1:1 weight ratio of $KMnO_4$ to sulfide was able to remove over 80% of the sulfide in 30 minutes. Increasing the ratio to 2:1 resulted in no detectable sulfide after 15 minutes. It should be noted that considerably more oxidant was needed here compared to the clean matrix case. This is due to the additional compounds found in spent caustic. These compounds exert a demand for the $KMnO_4$ which reduces its capability for direct sulfide oxidation.

Data Summary.

To put the data obtained into a general perspective, the concepts of first order rate constants and $t_{1/2}$ (half life) were emphasized. The data were transformed into terms of sulfide half life. The rate constants were calculated according to the following equation (*11*) and the rate constants (kA) were obtained by plotting ln [A]/[A$_0$] vs. time

$$\ln [A]/[A_0] = -kAt \qquad (4)$$

Where [A] = Sulfide concentration with time

kA = First Order Rate Constant

[A$_0$] = Sulfide concentration at time T$_0$

t = Time

Comparison of Tables I and II show the differences between $KMnO_4$ and MnO_2 for the oxidation of 10 mg/L sulfide in a clean matrix. As seen, the reaction with $KMnO_4$ is much faster than with MnO_2. As discussed, this was expected due to the dual oxidation pathway available when $KMnO_4$ is used (both direct and catalytic).

Table I. Sulfide Rate Constant and Half Life with $KMnO_4$:
Clean Water Matrix

$KMnO_4$: Sulfide (w:w)	Rate Constant $10^{-2}s^{-1}$	Half Life (Seconds)
1:1	1.90	14.0
2:1	3.60	8.0
4:1	5.60	2.0

Table II. Sulfide Rate Constant and Half Life with MnO_2:
Clean Water Matrix

MnO_2 : Sulfide (w:w)	Rate Constant $10^{-3}s^{-1}$	Half Life (Minutes)
1:1	1.46	6.0
2:1	2.26	2.5
4:1	4.42	2.0

The half life data and the rate constants for the spent caustic waste are presented in Table III. In this case, the initial sulfide concentration was 625 mg/L. As discussed previously, the increase in half life is due to the impurities in the caustic waste.

Table III. Sulfide Rate Constant and Half Life with KMnO₄: Spent Caustic Waste

$KMnO_4$: Sulfide (w:w)	Rate Constant $10^{-3}s^{-1}$	Half Life (Minutes)
0:1	0.22	90
0.04:1	0.50	30
0.16:1	0.72	15
0.96:1	1.47	10
1.90:1	4.02	5

The above information can be valuable when designing a treatment system. With the half life determined, equipment can be sized according to flow, retention time, chemical dose and degree of removal. Using the above case as an example, with a limited retention time of 30 minutes and a goal of 200 mg/L sulfide, a starting $KMnO_4$ dose would be about 60 mg/L (roughly a 0.1:1 weight ratio combined with adequate air and agitation). However, since all waste streams are unique in their characteristics, actual dosages have to be determined on a case by case basis.

Conclusions.

From this study, the following conclusions can be drawn:

1. In the presence of air, the amount of $KMnO_4$ needed for complete sulfide oxidation is less than theoretically calculated. Instead of the normally quoted value of 12:1, a more realistic and observed value is 4:1 or lower.

2. The MnO_2 formed from the initial oxidation acts as an oxygen transfer agent which can further oxidize sulfide. This was confirmed by using freshly precipitated MnO_2 to remove sulfide from water in the presence of air. It is this material that makes oxidation using less than the theoretical amount of $KMnO_4$ possible.

3. Low level use of $KMnO_4$ can, given sufficient time and air, effectively remove sulfide from spent caustic liquors.

Literature Cited.

1. *Chemical and Engineering News*. April 12, **1993,** p. 11.
2. 40 CFR Section 261.2 - 262.24; July 1, **1992.**
3. Stewart, R. *Oxidation by Permanganate*; Academic Press: New York, NY 1965; p. 28.
4. Eye, J. D.; Clement, D. P. "Oxidation of Sulfides in Tannery Waste Waters"; *Environmental Health Engineering*. University of Cincinnati, Cincinnati, OH, **1970.**
5. Narita, E.; Okabe, T. "The Formation and Some Properties of Hydrous Manganese (IV) Oxide"; *The Chem. Soc. of Japan.* **1980,** *vol.* 53, No. 2.

6. Arndt, D. *Manganese Compounds as Oxidizing Agents in Organic Chemistry;* Open Court Publishing Company: La Salle, IL. 1981; p. 28.
7. *Standard Methods for the Examination of Water and Wastewater;* Cleseceri, L.S.; Greenberg, A. E.; Trussel, R. R., Eds.; 17th Ed.; American Public Health Association: Washington, D.C., 1989.
8. Kotronaru, A.; Mills, G.; Hoffmann, M. R. "Oxidation of Hydrogen Sulfide in Aqueous Solution by Ultrasonic Irradiation"; *Env. Sci. & Tech.* **1992,** *vol.* 26, No. 12.
9. Cadena, F.; Peters, R. W. "Evaluation of Chemical Oxidizers for Hydrogen Sulfide Control"; *J. WPCF.* **1988,** *vol.* 60, No. 7.
10. Poole, B. "Pretreatment Systems For Tanners", Presented at The New England Water Pollution Control Association, Spring, **1980.**
11. Levine, I. N. *Physical Chemistry;* McGraw-Hill Book Company: 1978; pp. 481- 482.

RECEIVED March 14, 1995

Chapter 15

Photocatalytic Destruction of Atrazine Using TiO$_2$ Mesh

Kathleen C. Pugh, Douglas J. Kiserow[1], Jack M. Sullivan, and John H. Grinstead, Jr.

Environmental Research Center, Tennessee Valley Authority, P.O. Box 1010, Muscle Shoals, AL 35662–1010

The optimization of a technology for the TiO$_2$-mediated solar photocatalysis of atrazine is described. The target users of this technology might be agrochemical dealers, manufacturers, and possibly farmers. Studies were performed to determine the ideal cover material for such a reactor based on UV light transmitting properties. The best cover material was a UV transmitting acrylic. The TiO$_2$ employed for this technology was bound to fiberglass mesh. The effects of mesh amount, stirring, water impurities, concentration of pesticide, and source of UV light (mercury vapor lamp vs. solar) were also determined. The most efficient photocatalysis was achieved using five layers of mesh, a stirred reaction, water low in carbonate and other ions, a dilute waste stream, and solar irradiation rather than a mercury-vapor lamp. The formation and disappearance of eight intermediates were followed in a 2429 hour indoor experiment comparing pure and formulated atrazine. A modified reaction mechanism was proposed based on studies of the detected intermediates.

Over the last several years the Tennessee Valley Authority (TVA) has addressed environmental remediation problems through the research and development of innovative technologies. Our research emphasis has been on environmental problems related to agriculture. One such problem is the contamination of groundwater with pesticides.

Groundwater contamination via land application of dilute pesticide rinsates has been documented in several cases.[1,2]

[1]Current address: Army Research Office, Chemical and Biological Sciences Division, P.O. Box 12211, Research Triangle Park, NC 27709–2211

In an effort to provide an environmentally sound means of rinsate treatment, we have turned to advanced oxidative processes. Specifically, the applicability of TiO$_2$ photocatalysis for the destruction of pesticides in rinse or waste waters has been investigated.

The use of TiO$_2$ photocatalysis for pesticide rinsates offers several attractive features. First, the generation of pesticide rinsates is seasonal. During the 4-6 months of the growing season most of the rinsates would be produced. Therefore, a passive solar treatment technology might have as long as a year to destroy the rinsates accumulated during the previous growing season. Second, an agrochemical dealer typically generates a relatively low volume of rinsate waste (3,000-30,000 L/yr/site). These low volumes suggest the treatment apparatus need not be exceedingly large and may therefore be of relatively low cost. Third, TiO$_2$ is a nontoxic, relatively stable semiconductor, allowing long-term use of a single batch of TiO$_2$ and safe disposal of "spent" catalyst.

The target users of this technology might be agrochemical dealers, manufacturers, and possibly farmers. Several features of the solar photocatalytic system would allow for easy, inexpensive operation by the user. First, the use of TiO$_2$ impregnated mesh rather than a TiO$_2$ slurry removes the need for filtration of the remediated rinsewaters. In addition, the passive solar technology under development requires no mirrors or lamps, which should eliminate the need for adjustments as well as keep the capital cost and operations costs low.

Photocatalysis using TiO$_2$ has been shown to significantly degrade or mineralize pesticides in aqueous solution.[3] The pesticide selected for use during development of this technology was atrazine. Atrazine, an *s*-triazine, is one of the most widely used herbicides in the United States, especially in the Midwest. Triazines are used as herbicides for crops such as corn, sorghum, and sugar cane as well as to control broadleaf and grassy weeds along railways, highways, and on rangeland.[4]

Several parameters have been investigated during the research and development phase of this work. Studies have been carried out to determine the best cover material, the optimum number of layers of mesh, the effect of mixing, the effect of water quality on degradation rate, comparisons of photocatalysis by mercury lamp vs. solar photocatalysis, and the effect of concentration on degradation rate.

In addition to optimizing the system toward the most efficient degradation of atrazine, particular attention was given to which intermediates and end products were generated during this process. In order to obtain permits for this remediation technology, it is likely the identity and toxicity of intermediates and end products will have to be furnished to the appropriate authorities.

Observations of intermediates generated during degradation experiments have led to the proposal of a degradation mechanism somewhat different than one proposed earlier by Pelizzetti, et al.[3] A previously unreported intermediate, 2-amino-4-hydroxy-6-isopropylamino-*s*-triazine (OAIT), has been detected and appears to play a significant role in the degradation of atrazine under the conditions reported here.

Experimental

The nomenclature system of Cook,[5] Adams,[4] and Hapeman-Somich[6] is used here: A, amino; C, chloro; E, ethylamino; I, isopropylamino; O, hydroxy; T, triazine ring. Another abbreviation includes M, methoxy.

Photocatalytic degradation of atrazine and byproduct formation were monitored using GC and HPLC. GC measurements were made using a Varian 3600 gas chromatograph equipped with a SPB-5 megabore column and a N/P detector. HPLC measurements were made using a Beckman Model 126 System Gold HPLC equipped with a Model 126 Programmable Solvent Module, a Model 168 Diode Array Detector Module, and a Model 507 Autosampler with a 100 µL sample loop. HPLC analysis was performed using either a Merck LiChrosorb RP-C18 (250 mm x 4.6 mm ID) or a Shandon Hypercarb S graphitized carbon column (100 mm x 4.6 mm ID). The column was preceded by a guard cartridge containing identical packing material (Alltech, direct connect guard cartridge, 10 mm x 4.6 mm). UV absorbance was monitored at 220 nm.

Three HPLC methods were used for intermediate analysis. Method 1 was used to identify atrazine, OAIT, CEAT, CIAT, CAAT, OIET, and MEIT. Method 2 was used to identify OOOT, OOAT, OAAT, and CAAT, although OOOT and OAAT coeluted and OOAT and CAAT coeluted. Method 3 was used to identify OAAT only (OAIT could be detected by this method as well).

HPLC Method 1: Merck Lichrosorb RP C18 Column, gradient: 40% B for 11 min, 40-70% B in 10 min, 70% B for 24 min, 70-0% B in 15 min, 0-40% B in 5 min, where A = 50 mM ammonium acetate in water, pH 7.4 and B = 50 mM ammonium acetate in methanol, flow = 0.5 mL/min, 220 nm detection. Species Detected: CAAT (7.80 min), OAIT (9.16 min), CEAT, (16.18 min), CIAT (25.28 min), OIET (27.12 min) MEIT (34.75 min), atrazine CIET (35.87 min). HPLC Method 2: Shandon Hypercarb S graphitized carbon column, isocratic gradient, 30% methanol/water, flow = 1.0 mL/min, 220 nm detection. Species detected: OAAT (6.04 min), OAIT (10.57 min). HPLC Method 3: Merck Lichrosorb RP C18 Column, isocratic gradient, 100 mM potassium phosphate buffer, pH 6.7, flow 0.5 mL/min, 220 nm detection. Species detected: CAAT and OOAT (coelute, 7.80 min); OAAT and OOOT (coelute, 9.10 min).

Experimental samples were prepared using formulated atrazine (40.8% atrazine, atrazine 4L Herbicide from St. Augustine Turf, Southern Agricultural Insecticides, Inc.), pure atrazine (98% atrazine, Supelco), CAAT, and OAIT (each 97% pure, provided at no cost by Ciba-Geigy Corporation). Standards for HPLC and GC analysis were prepared from the previously described chemicals as well as CIAT and CEAT (99% each, Chem Services, West Chester, PA), OAAT, and OOAT (94% and 98%, respectively, provided at no cost by Ciba-Geigy) and OOOT (98%, Aldrich Chemical Co.).

TiO_2 impregnated fiberglass mesh was used as the catalyst (Nutech Environmental, London, Ontario).

For both indoor and outdoor experiments, UV light in the 290-385 nm range was measured with an Eppley Laboratory Ultraviolet Radiometer (No. 29403). In the indoor experiments TiO_2 was irradiated with a Canrad-Hanovia 450-W high-pressure mercury-vapor lamp (Model 679A-0360) located in a Pyrex sleeve. A pyrex sleeve was placed around the mercury vapor lamp to filter light below 280 nm so that it would more closely approximate solar irradiation. Uniform mixing was achieved by magnetic stir plates located under the experimental box. The experimental apparatus is shown in Figure 1. Experiments were carried out in which approximately 1 L of aqueous pesticide solution (993.00 g) was subjected to irradiation by the mercury lamp. Samples could be removed by pipette through the side arms without turning off the lamp.

Outdoor experiments were conducted on the roof of TVA's National Environmental Research Center building in Muscle Shoals, Alabama. The outdoor experimental reactors are shown in Figure 2. Rectangular 41 x 33 x 2.7 cm 316 stainless steel reactors were slanted at an angle of 34 degrees to the horizontal, facing south. The cover plate for the reactors was 0.32 cm clear acrylic plastic. Three layers of TiO_2 mesh were mounted on a 30.5 x 30.5 cm glass frame designed to slide in from the top of the reactor. Each reactor contained 2800 g of atrazine solution which was mixed by sparging from the bottom with air. Samples were taken with pipettes through a loose fitting threaded orifice-plug assembly located above the liquid level. The top of each reactor was loosely taped with cellophane to reduce evaporation losses.

UV absorbance measurements were made using a Varian DMS 90 UV-visible spectrophotometer coupled to an IBM XT personal computer.

Water from several locations used to prepare atrazine samples was tested for purity by Biotransformations Incorporated, Colorado Springs, Colorado.

Results And Discussion

Reactor Cover Materials. The reactor cover material for a solar TiO_2 photocatalytic system is critical because of the nature of the solar spectrum and the characteristic absorbance range of TiO_2. Only about 1% of the solar spectrum is absorbed by TiO_2,[7] thus maximization of the efficiency of the photocatalytic process requires, in part, maximization of the UV light irradiating TiO_2.

Figure 3 shows the UV transmission of several materials which are representative of those tested. The wavelength range of interest was primarily in the 300-380 nm range since the solar spectrum has little output below 300 nm and TiO_2 does not absorb above 380 nm. In this range, the order of UV transmission was: UV transmitting acrylic > Pyrex > window glass > UV absorbing acrylic. The implication is that atrazine degradation should be most effective in the same order.

Pseudo-first-order kinetics have been reported for atrazine degradation under a variety of conditions [3,4,8] and for the degradation of other molecules such as trichloroethylene and diisopropyl ether.[9,10] The basic kinetic scheme for

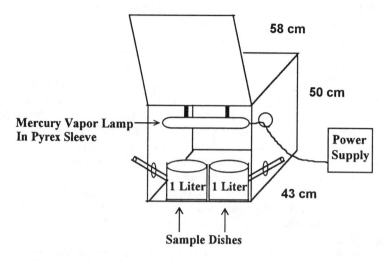

Figure 1. Indoor irradiation apparatus with mercury-vapor lamp.

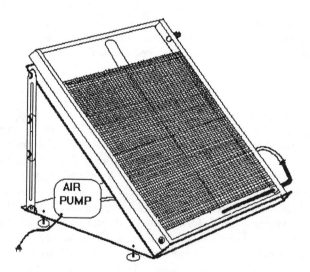

Figure 2. Solar reactor.

pseudo-first-order kinetics appears to work well for the system studied here, under the chosen conditions.

Atrazine photocatalysis was carried out using the materials described in Figure 3 as reactor covers. The rate constant was determined from the slope of the plot of $\ln([C]/[C]_0)$ (where [C] is the concentration of atrazine) vs. time, shown in Figure 4. The coefficients of determination, r^2, are quite high (0.997 or greater) indicating good fits. Holding all other variables constant, it was observed that the three UV-transmitting covers all yielded approximately the same degradation rates. Because the acrylic is considerably less expensive than Pyrex (ca. \$15.00/m^2 vs. \$98.00/m^2) and may be more durable, it is potentially the best cover for use in a large-scale reactor.

It should be emphasized that although the plotted data show that the degradation of atrazine, under specific conditions, appears to follow pseudo-first-order kinetics, this is not proof of a specific mechanism. Variation in parameters such as initial concentration, quantity of TiO$_2$ mesh, UV photon flux, etc., may all affect the rate of atrazine degradation. To make valid comparisons between the data presented herein, all variables were held constant except the one of interest.

TiO$_2$ Mesh. The rate and efficiency of atrazine degradation will depend to a large extent on the quantity of TiO$_2$ used. Since immobilized TiO$_2$ (in a mesh) is desirable for this work, it is clear that for a specific number of layers of mesh, UV light will no longer reach each layer due to shielding by the others.

The transmission of UV light through multiple layers of mesh was measured by locating the mesh between the mercury-vapor lamp and a radiometer and measuring the intensity in the 290-385 nm range. Figure 5 shows that as each layer of mesh was added, the UV intensity decreased. When six to eight layers of mesh were used, there were only small changes in intensity. When nine layers were in place no UV light was detected. This data indicates that five layers of mesh is probably the optimum number with respect to achieving maximum photocatalytic activity without employing excess mesh.

This hypothesis was tested by degrading atrazine with one, three, five, and seven layers of mesh. The results (Figure 6) correspond to those in Figure 5 and show that the rate constant increased until five layers were used. Presumably, any additional layers were totally screened from UV light and no additional measurable photocatalytic activity would be observed. These experimental results showing five layers of mesh to be optimal will be applied to the design of a prototype unit for this technology.

Effect of Stirring on the Photocatalytic Rate. One of the basic concerns with respect to a viable photocatalytic technology for rinsate treatment is the effect of stirring on the degradation rate of a waste solution. If efficient photocatalytic degradation is not possible without mechanical stirring, this must be taken into consideration during the design of the photocatalytic reactor.

Figure 7 shows the results of atrazine degradation with and without stirring. The data are presented in Figure 7 plotted as $\ln([C]/[C]_0)$ vs. time. The

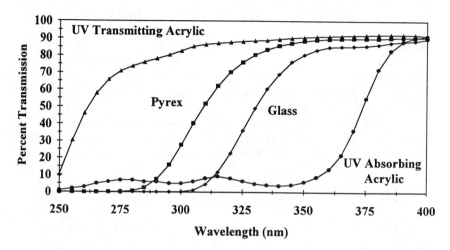

Figure 3. UV transmission of different reactor cover materials.

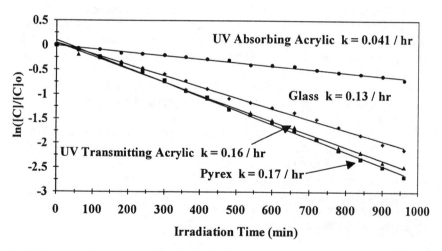

Figure 4. Best fit of $\ln([C]/[C]_0) = -kt$ to data from atrazine degradation using different reactor covers.

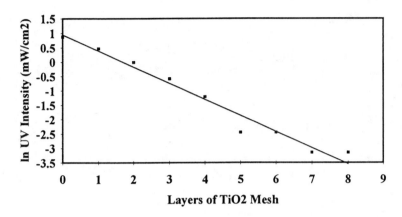

Figure 5. The ln of UV light intensity vs. numbers of layers of TiO2 mesh.

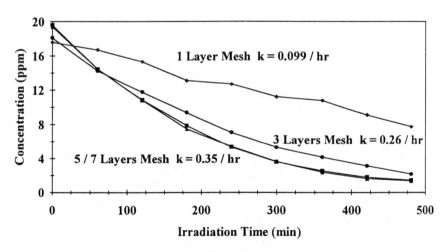

Figure 6. Atrazine degradation using different numbers of layers of TiO2 mesh.

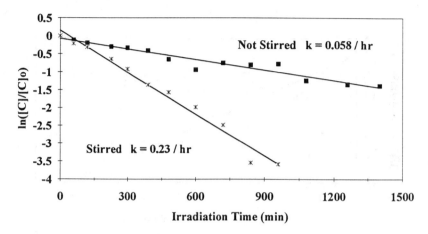

Figure 7. Best fit of ln([C]/[C]$_0$) as a function of time for identical samples photocatalytically degraded with and without stirring.

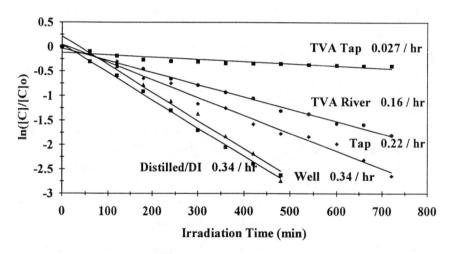

Figure 8. Best fit of ln([C]/[C]$_0$) = -kt to data from atrazine degradation using water from different sources.

rate of atrazine degradation without stirring is 0.058 hr^{-1} and with stirring is 0.23 hr^{-1}. The difference is approximately a factor of 4.

Stirring during photocatalysis may be necessary to bring unreacted atrazine molecules in contact with hydroxyl radicals present at the TiO$_2$ surface. The difference in rate with and without stirring is particularly important, considering that the volume of solution in this test was about 1 L and a working unit may be designed to hold several hundred times this amount.

Effect of Water Quality on Degradation Rate. A potentially critical parameter in the degradation process is the type of water in which the pesticide waste is dissolved. Water from different locations may have widely varying properties due to contaminants from chemical treatment and natural processes. The experimental results reported above have been presented for studies carried out in distilled and deionized water, which appears to be pure from independent test results. Although this is the best water to use for experiments with respect to limiting the introduction of impurities, it may not be comparable to the water used to prepare actual pesticide rinsates.

The photocatalytic degradation of atrazine was carried out in water from five different northwest Alabama locations to investigate the effect of each water sample on the photocatalytic process. The water samples were: (1) tap water from the city of Florence, (2) tap water from TVA, Muscle Shoals, which is treated at the TVA treatment plant, (3) Tennessee River water from TVA, Muscle Shoals, which is untreated, (4) well water from Florence, and (5) distilled and deionized water prepared at TVA, Muscle Shoals. The degradation of atrazine dissolved in these waters as a function of time is shown in Figure 8. The data were fit to obtain the rate constants (given in the figure) for quantitative comparison. The rate of atrazine degradation, from slowest to fastest, was: TVA treated tap water, TVA untreated river water, Florence city tap water, Florence city well water, and TVA distilled and deionized water.

Several interesting observations may be made with respect to Figure 8. First, the water with the poorest photocatalytic activity was TVA tap water. All atrazine degradation was somewhat similar in rate except for this case. Second, well water and distilled/deionized water gave very similar results, indicating that the natural impurities present in this well water have no significant effects on photocatalysis. Third, the degradation in Florence city tap water was also quite efficient. This is somewhat surprising since one might expect this water to be similar to TVA tap water in that treated water, in general, would have similar impurities which would be detrimental to the photocatalytic process. The difference in water may be due to the fact that Florence tap water comes from Cypress Creek while TVA tap water comes from the Tennessee River, or to differences in the treatment methods at the two locations. It is important to note that although the rate of atrazine degradation using TVA tap water was quite slow in this indoor study (Figure 8), experiments conducted outdoors using small-scale prototype reactors and TVA tap water showed considerably faster rates of atrazine degradation, probably due to the increased intensity of UV light in the wavelength

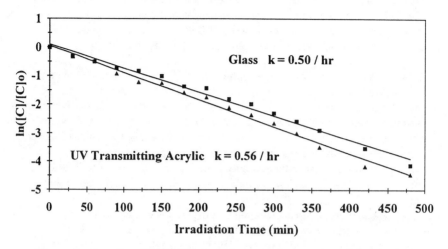

Figure 9.   Best fit of $\ln([C]/[C]_0) = -kt$ to data from atrazine degradation outdoors.

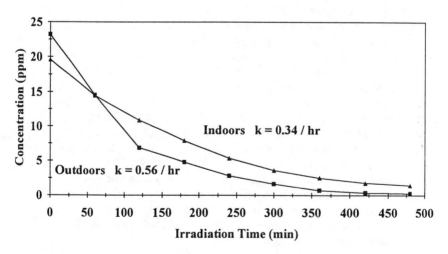

Figure 10. Comparison of atrazine degradation indoors and outdoors.

range of interest (see discussion below). The results imply that even in the water with the highest concentration of impurities (TVA tap water), atrazine degradation may be carried out effectively under solar irradiation.

A standard water analysis was performed by Biotransformations Incorporated and gives some insight into which water impurities may affect the degradation of atrazine. The decrease in the rate of atrazine degradation is concurrent with an increase in ions, including $CaCO_3$, in solution. Carbonate ion is a hydroxyl radical scavenger, thus it is not a surprise that the rate of photocatalysis decreased with increasing $CaCO_3$ concentration. Future work may involve a detailed study of the effects of commonly found impurities and how to overcome their inhibition of photocatalysis.

Indoor vs. Outdoor Degradation Rate. Two photocatalytic reactors were constructed for use under solar irradiation in which the sides and top were either glass or the UV transmitting acrylic. The construction allowed sunlight to irradiate the mesh at all times from any direction while located on the roof of the research center. The outdoor experiments used 5.0 g of mesh and 993.00 g of formulated atrazine in each reactor, which were the conditions employed for indoor experiments. The configuration of the outdoor reactors was very similar to that used indoors, thus any difference in photocatalytic rate would be due primarily to the use of solar irradiation as opposed to the mercury-vapor lamp. The reactors were placed on the research center roof on a sunny to partly sunny day with the sun almost directly overhead at 12 noon.

Figure 9 shows the degradation of atrazine as a function of time for the UV transmitting acrylic and the glass reactors exposed to solar radiation. The rate of degradation is similar for both reactors. The rate constants for the glass and the UV transmitting acrylic reactors were determined and were within approximately 10% of each other in spite of significant differences in the UV transmission.

The rate of atrazine degradation outdoors ($k = 0.56$ h^{-1} using the UV transmitting acrylic, Figure 10) was significantly faster than for a similar experiment carried out indoors ($k = 0.34$ h^{-1}, Figure 10). This was probably a result of the higher UV intensity outdoors. The solar intensity in the range 290-385 nm was measured every half hour on the day of the experiment and the average was 3.2 mW/cm^2 while the average intensity for the UV lamp after reaching a steady state during an experiment is 2.2 mW/cm^2. The percentage of difference between the UV intensity outdoors and indoors roughly corresponds to the percentage of difference between the rates. This indicates that during the summer months, a faster rate of atrazine degradation may be expected during outdoor photocatalytic experiments.

Effect of Concentration on Degradation Rate. Because the importance of substrate preadsorption on TiO$_2$ may be determined using a modified Langmuir-Hinshelwood (LH) kinetic model,[11,12,13,14] the degradation of atrazine using solar irradiated TiO$_2$ at varying concentrations of atrazine (20.7, 14.3, 6.9, and 3.3 ppm) in TVA tap water was investigated. Three layers of TiO$_2$ mesh were used

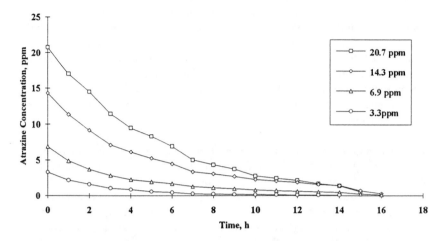

Figure 11. Destruction of atrazine at various concentrations as a function of time. Experiments were carried out in TVA tap water with three layers of mesh using solar irradiation.

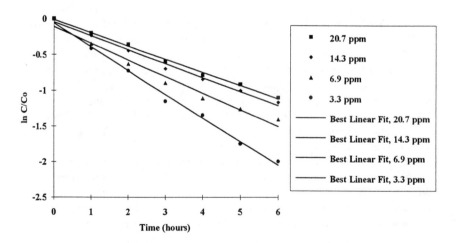

Figure 12. Pseudo-first-order rate plot of ln C/Co vs. time for the solar destruction of atrazine.

for this experiment. The kinetic data were derived from six samples taken at one-hour intervals during the first day. The sky was partially overcast resulting in UV light (295-385 nm) readings ranging from 1.1 mW cm^{-2} to 3.4 mW cm^{-2}.

The destruction of atrazine as a function of time is shown in Figure 11. Despite the variable irradiation, linear first-order rate plots were obtained in each case, resulting in pseudo-first-order rate constants of 0.184 h^{-1}, 0.194 h^{-1}, 0.232 h^{-1}, and 0.331 h^{-1}, for atrazine concentrations of 20.7, 14.3, 6.9, and 3.3 ppm, respectively (Figure 12). Hence, there is an increase in rate constant with decreasing initial atrazine concentration.

The data from this experiment were plotted using the Langmuir-Hinshelwood equation,

$$1/R_0 = 1/ k'KC_0 + 1/k'$$

where R_0 is the initial rate of reaction, C_0 is the initial atrazine concentration, K is the adsorption constant, and k' is the reaction rate constant (Figure 13). A reaction rate constant, k', of 6.45 ppm h^{-1} and an adsorption constant, K, of 0.0643 ppm^{-1} were derived from the LH plot. The inference has been made that a linear relationship of an LH plot is indicative of photocatalytic decomposition occurring completely on the catalyst surface,[15] although Turchi and Ollis have disputed this interpretation.[11]

Mechanism

Most of the work to this point has focused only on the degradation of atrazine. However, the ultimate goal of this photocatalytic process is the complete mineralization of all intermediates. To track the progress toward this goal, the intermediates formed during photocatalysis must be monitored. Furthermore, it will probably be necessary to provide identification and toxicity data for the intermediates generated to obtain permits for this process from the appropriate authorities.

In the case of TiO$_2$ degradation of atrazine, mineralization has not been observed to occur under the conditions described above nor under conditions described by Pelizzetti, et al.[3] Instead, the end product commonly observed has been cyanuric acid (OOOT). Under the conditions described above, the other intermediates identified to this point were OAIT, CIAT, CEAT, CAAT, OAAT, OOAT, and OIET. Although it was possible to detect MEIT, this was not an observed intermediate of the photocatalytic process. A long-term experiment comparing pure and formulated atrazine using five layers (5.0 g) of mesh and 993.00 g of each type of atrazine solution was conducted in the indoor experimental apparatus (Figure 1). The experiment was carried on for 2429 h of irradiation, which would correspond to approximately 303 eight-hour days or 10 months. The photocatalytic degradation was followed by HPLC, with GC used to confirm analyses when possible. It was not possible to detect any of the hydroxy-containing intermediates by GC, so HPLC was used exclusively for the detection of those intermediates. Figures 14-16 show examples of the HPLC chromatograms from the three methods used for HPLC analyses.

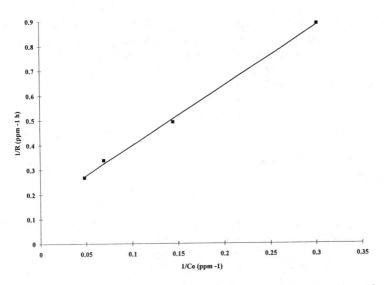

Figure 13. Langmuir-Hinshelwood plots for the solar destruction of atrazine in TVA tap water.

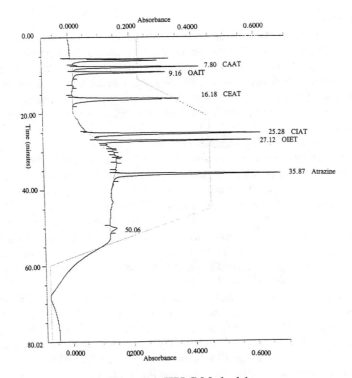

Figure 14. HPLC Method 1.

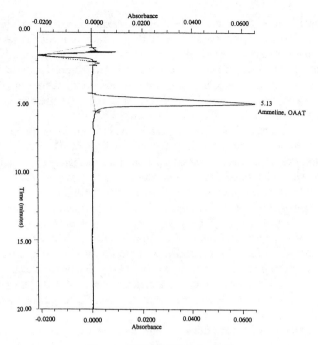

Figure 15. HPLC Method 2.

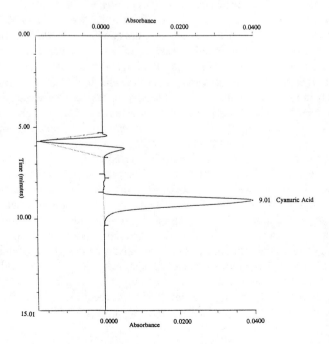

Figure 16 HPLC Method 3.

Table I shows the results of the long-term experiment. Intermediates were identified by coelution with standards. Although eight intermediates have been identified in this manner, several HPLC and GC peaks remain unidentified at this point. The time at which an intermediate was detected (Appearance), the time at which the intermediate peaked in concentration (Peak), and the time at which the intermediate could no longer be detected (Disappearance) are listed in Table I. Most intermediates could be detected to at least 0.1 ppm.

Several interesting observations can be made from the long-term experiment. First, a previously unreported intermediate, OAIT, was detected. In earlier work, Pelizzetti, et.al. proposed a degradation mechanism in which OIET was detected as an intermediate, but was thought to comprise less than 10% of the initial degradation of atrazine due to lack of chloride formation in the early stages of degradation.[3] However, OAIT appears to be a major intermediate under the conditions described here, accounting for at least as much of the degradation pathway to cyanuric acid as CAAT. Furthermore, OIET, which was detected as both a contaminant of the starting atrazine solutions as well as an intermediate, did not behave as a precursor intermediate for OAIT. The peak concentration of OIET never exceeded one-fourth the peak concentration of OAIT. Based on these unique observations, we have suggested the modified degradation pathway shown in Figure 17.

A second set of observations from the long-term experiment are the differences in the rates of formation and disappearance of the intermediates from the pure and formulated reactions. The disappearance of atrazine was slower by a factor of 1.07 for the formulated experiment as compared to pure atrazine. The time at which most intermediates appeared was identical, except for OOOT, which appeared 3.3 times later in the case of pure atrazine. The time at which the intermediates peaked in concentration was generally later for the formulated case, although an exception was noted for CAAT, which peaked 1.86 times later in the pure case. Intermediates also generally took somewhat longer to disappear completely in the formulated case. These results suggest that the surfactants which are used to make formulated atrazine slow the general rate of photocatalytic degradation by interfering with the catalyst, possibly through adsorption or simply by competing with atrazine for oxidation by the catalyst.

However, the slower formation of CAAT and OOOT in the case of pure atrazine may have mechanistic implications. The presence of surfactants, which may act as hydroxyl radical scavengers/competitors, appears to increase the rate of formation of CAAT and OOOT. When Glaze et al.[16] introduced a hydroxyl radical scavenger during TiO_2 photodegradation of tetrachlorethylene, the rate of formation of one intermediate, dichloroacetic acid, increased. They concluded that a dual mechanism occurred during photocatalysis involving both oxidation and reduction. Similarly, atrazine photodegradation by TiO_2 may proceed by a dual mechanism of both oxidation and reduction. This hypothesis is supported by the observation that ozonation of atrazine, which proceeds through hydroxyl radical oxidation,[4] does not produce any oxidation products of CAAT.[4,6]

Experiments are underway to further explore the hypothesis that reduction may play a role in TiO$_2$ photodegradation of atrazine intermediates.

Because OAIT has not been reported previously as a TiO$_2$ photocatalytic degradation product of atrazine, an experiment was performed to try to establish its place in the degradation pathway. A 993.00 g solution of 13 ppm OAIT was irradiated with the mercury-vapor lamp using five layers (5.0 g) of TiO$_2$ mesh. The only detected degradation product of this experiment was OAAT (Figure 18). This observation lead to the proposal of the modified degradation pathway in Figure 17. Experiments are continuing to determine whether CIAT might also contribute to the formation of OAIT.

Summary And Conclusions

The results of the optimization studies described above suggest that a reactor covered with UV transmitting acrylic equipped with five layers of TiO$_2$ mesh in which the waste water is stirred would function most efficiently. It also appears that solar radiation is more effective than a mercury-vapor lamp. Experiments using various concentrations of atrazine showed that the more dilute the waste stream, the faster the degradation of atrazine. Water impurities, especially radical scavengers such as carbonate ion, caused slower rates of degradation than when little or no impurities were present. However, comparable rates of atrazine degradation have been observed when solutions high in carbonate ion undergo solar irradiation as opposed to irradiation with the mercury-vapor lamp.[17] Presumably this was because solar UV intensities are significantly higher than those emitted by the mercury-vapor lamp.

The relative rates of intermediate degradation were determined from the long-term indoor experiment. Whereas formulated atrazine was fully degraded in 8 h, CAAT required 2.4 months and OAAT required 2.7 months for complete degradation. The end product, OOOT, reached a plateau in concentration after 5.1 months. Although the intermediates required orders of magnitude more time to degrade than atrazine itself, it should be noted that these results are based on irradiation with a mercury-vapor lamp rather than solar irradiation. In addition, estimated degradation times were based conservatively on "8-hour days." Based on the outdoor (solar) experiments completed to date, it appears that much faster rates of degradation are likely under summer solar irradiation conditions when UV intensity and day length are at a yearly peak. One might expect the times required for degradation to be as little as half those reported for the indoor experiment.

The mechanistic studies described here have lead to the proposal of a slightly different mechanism in which OAIT is a major degradation product. Formulated atrazine was found to degrade 7% more slowly than pure atrazine, although certain intermediates (CAAT, OOOT) form more quickly in the prescence of the formulations. This may be indicative of a dual mechanism in which both oxidation and reduction are catalyzed by TiO$_2$.

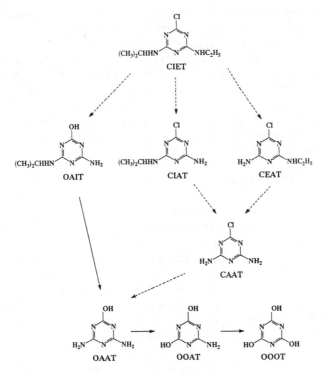

Figure 17. Degradation mechanism for atrazine photocatalysis.

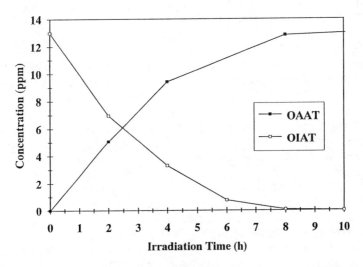

Figure 18. Degradation of pure OAIT and formation of OAAT.

The observed end product of the experiments described here was OOOT (cyanuric acid) rather than mineralization. Cyanuric acid is considered to be practically non-toxic.[18,19] Therefore, the transformation of atrazine to cyanuric acid (OOOT) represents a marked decrease in toxicity of the pesticide rinsate. Future experiments will be aimed at determining conditions under which the intermediates are degraded faster and, if possible, OOOT is mineralized.

Table I
Time (In Hours) For The Appearance, Peak In Concentration, And Disappearance Of Atrazine And Its Intermediates

Intermediate	Appearance		Peak	Disappearance
Atrazine (CIET)	Pure	-----	-----	7.33
	Form.	-----	-----	7.83
OAIT	Pure	0.5	11.83	103.8
	Form.	0.5	25.83	127.8
CIAT	Pure	1	4.5	9.83
	Form.	1	5.5	11.83
CEAT	Pure	0.5	2.5	9.83
	Form.	0.5	5	9.83
CAAT	Pure	2	103.8	345
	Form.	2	55.83	584
OAAT	Pure	0.5	31.83	512
	Form.	2	31.83	656
OOOT	Pure.	5	1095.6	-----
	Form.	1.5	1216	-----

NOTATIONS

C18, silicon support modifed with 18 carbon chains, CAAT, 2-chloro-4,6-diamino-*s*-triazine,CEAT, 2-amino-4-chloro-6-ethylamino-*s*-triazine,CIAT, 2-amino-4-chloro-6-isopropylamino-*s*-triazine, CIET, atrazine, 2-chloro-4-ethylamino-6-isopropylamino-*s*-triazine, GC, gas chromatography, HPLC, high performance liquid chromatography, MEIT, atraton, 2-methoxyatrazine, OAAT, ammeline, 2,4-diamino-6-hydroxy-*s*-triazine, OAIT, 2-amino-4-hydroxy-6-isopropylamino-*s*-triazine, OOAT, ammelide, 2-amino-4,6-dihydroxy-*s*-triazine, OOOT, cyanuric acid, 2,4,6-trihydroxy-*s*-triazine, RP, reverse phase.

REFERENCES

‡ Current address: Army Research Office, Chemical and Biological Sciences Division, P.O. Box 12211, Research Triangle Park, NC 27709-2211

[1.] Aharonson, N. *Pure Appl. Chem.* **1987**, *59*, 1419.

[2.] Parsons, D. W.; Witt, J. M. Oregon State University Extention Service Report of a 1988 Survey of State Lead Agencies, Corvallis, Oregon, **1988**.

[3.] Pelizzetti, E.; Maurino, V.; Minero, C.; Carlin, V.; Pramauro, E.; Zerbinati, O.; Tosato, M. L. *Environ. Sci. Technol.* **1990**, *24*, 1559.

[4.] Adams, C. D.; Randtke, S. J. *Environ. Sci. Technol.* **1992**, *26*, 2218.

[5.] Cook, A. M.; Beilstein, P.; Grossenbacher, H.; Hutter, R. *Biochem. J.* **1985**, *231*, 25-30.

[6.] Hapeman-Somich, C. J.; Gui-Ming, Z.; Lusby, W. R.; Muldoon, M. T.; Waters, R. *J. Agric. Food Chem.* **1992**, *40*, 2294.

[7.] Gerischer, H.; Heller, A. *J. Electrochem. Soc.* **1992**, *139*, 113.

[8.] Kolpin, D.; Kalkoff, S. *Environ. Sci. Technol.* **1993**, *17*, 134.

[9.] Glaze, W.; Kenneke, J.; Ferry, J. *Environ Sci Technol* **1993**, *27*, 177.

[10.] Wallington, T.; Andino, J.; Potts, A.; Rudy, S.; Siegl, W. *Environ. Sci. Technol.* **1993**, *27*, 98.

[11.] Turchi, C. S.; Ollis, D. F. *J. Catal.* **1990**, *122*, 178.

[12.] Jenny, B.; Pichat, P. *Langmuir* **1991**, *7*, 947.

[13.] Langmuir, I. *Trans. Faraday Soc.* **1921**, *17*, 621.

[14.] Matthews, R. W. *J. Chem. Soc., Faraday Trans. I* **1989**, *85*, 1291.

[15.] Matthews, R. W. *Water Res.* **1990**, *24*, 653.

[16.] Glaze, W. H., Kenneke, J. F., Ferry, J. L. *Environ. Sci. Technol.* **1993**, *27*, 177-184.

[17.] Unpublished observation, Jack Sullivan.

[18.] Hammond, B.G.; Barbee, S.J.; Inoue, T.; Ishida, N.; Levinskas, G.J.; Stevens, M.W.; Wheeler, A.G.; Cascieri, T. "A Review of Toxicology Studies on Cyanurate and its Chlorinated Derivatives," *Environmental Health Perspectives*, **1986**, 69, 287-292.

[19.] Pugh, K.C. "Toxicity of Atrazine and its Degradation Products" TVA Y Bulletin No. 235, December, **1993**.

RECEIVED June 8, 1995

Extraction and Precipitation

Chapter 16

Bench-Scale Chemical Treatability Study of the Berkeley Pit Water

Hsin-Hsiung Huang and Qi Liu

Department of Metallurgical Engineering, Montana Tech of the University of Montana, Butte, MT 59701

Lime and limestone are the most commonly used and most effective chemical reagents to treat acid mine drainage. They can be used to treat the Berkeley Pit water in Butte, Montana by two-stage neutralization processes to meet EPA Water Quality Criteria for Aquatic Life (Gold Book). Over 92% of copper in the Berkeley Pit water can also be cemented by industrial scrap iron without any ill effect on the subsequent neutralization treatment. The two-stage lime neutralization process, with several polishing steps tested, was confirmed by an 80 gallon water test.

The oxidation and leaching of sulfide minerals in the mines and in mine wastes are the major causes of producing acid mine drainage. The Berkeley Pit, located at the northeast edge of Butte, Montana, is one of the world's largest ore deposits that contains copper and other metals. Mining began in the Butte area in the late nineteenth century. The Anaconda Copper Mining Company began open pit mining in 1955, and during the operation, drainage water was pumped, treated and discharged. Since the shut-down of the Berkeley Pit in 1982, the drainage pumps were turned off. Both underground mine workings and the Pit are flooding with the water.

In terms of contained volume of the water and quantity of metal pollutants, the Berkeley Pit is unmatched by any acid producing mine in the United States and possibly in the world. It has been filling at a rate that ranges from 5 to 7.6 million gallons per day, and has accumulated over 20 billion gallons. The water will probably reach an alluvium bedrock contact exposed in the Pit wall approximately in the year 2011. This will create a serious threat to the ground water quality in the Butte area. About 1/3 of the water entering the Pit is surface water and 2/3 is from underground water.

0097–6156/95/0607–0196$12.00/0

The hazardous constituents in the water are several orders of magnitude greater than discharge regulations and they clearly need removal and disposal. To develop or to evaluate processes for treating the water requires an understanding of the chemical properties and the sources of the water itself.

CHEMISTRY

The chemistry of the Berkeley Pit water was studied for treatment purposes. Unlike most of the waters which are anionically dominated by chloride or carbonate, the Pit water is dominated by sulfate. Sulfate waters are usually associated with metals or coal mining wastes or with current geothermal activity. The Berkeley Pit water also contains extremely high levels of dissolved heavy and transition metals, and is highly oxidized and acidified compared even to most of the acid mine drainage. The major cation is iron. The chemistry study for speciation calculation, saturation with solids, neutralization and oxidation are available from *(1)*.

Composition of the Berkeley Pit Water

The Berkeley Pit water has been sampled and reported on by EPA, Montana Bureau of Mines and Geology (MBMG), and Montana Tech since 1984. The composition of the water about 200 feet below the surface, the EPA Gold Book regulations and the drinking water standards are listed in Table 1.

Table 1. Composition of the Berkeley Pit Water and EPA Water Criteria (ppm)

	Nov. 84 62 feet	Jun. 85 100 feet	Oct. 86 200 feet	Oct. 87 216 feet	May. 91 225 feet	Oct. 92 200 feet	Gold Book[a]	Drinking Water
[Al]	142	172	192	193	288	304.5	0.087	/
[As]	0.2	0.43	0.04	1.2	0.83	0.43	0.05	0.05
[Cd]	1.54	1.62	1.74	1.76	1.57	2.00	0.0023	0.01
[Ca]	477	435	457	479	492	525.1	/	/
[Cu]	164	229	204	202	191	215.1	0.042	1
[Fe]	256	451	918	1010	1088	1112	1	0.3
[K]	4.4	8.8	24.3	18.7	20	19.9	/	/
[Mg]	236	261	291	279	418	517	/	/
[Mn]	106	116	144	161	182	225.5	/	0.05
[Ni]	/	/	0.91	0.99	1.05	0.91	/	/
[Na]	61.7	60	65.8	70.5	68	107.7	/	/
[Pb]	/	/	/	0.66	0.08	<0.13	0.01	0.05
[Zn]	255	329	460	494	552	636.6	0.23	/
SiO_2	/	/	/	/	/	51.24	/	/
Cl^-	12.3	8.3	/	21.8	10.9	/	/	250
Sulfate	4410	5550	/	6940	8010	7700	/	250
pH	2.78	2.48	/	3.15	2.84	2.96	/	6.8-8.5
Eh, mv	/	/	/	463	650	621	/	/

[a]EPA Gold Book: Quality Criteria for Water, 1986 and revision 1991 based on the hardness of 250 ppm for aquatic life.

TREATABILITY

Considering the water in the Berkeley Pit and water used in the nearby concentrator, Atlantic Richfield Company (ARCO) proposed ten flow options for treating the Berkeley Pit water (2). To evaluate each flow option, a bench scale treatability program was developed. The tests included in-situ and external treatments of the water. The in-situ tests simulated the pre treatment processes by adding chemicals directly into the water from the Berkeley Pit. Chemicals used include: lime, caustic soda, magnesium hydroxide, potash, potassium sulfate, and tailings slurry from milling. The external tests evaluated pump-treat-discharge processes by studying lime precipitation and sulfide precipitation. The external pre treatment included copper cementation.

This paper describes the development of the neutralization process for treating the Berkeley Pit water in order to identify low cost reagents and to meet the EPA Gold Book standards. The tests included

- One-stage neutralization;
- two-stage neutralization;
- large scale two-stage neutralization, and
- copper cementation prior to neutralization

The results from other bench scale tests are available from References (3), (4).

General Test Procedure

Most of the bench scale tests were performed at room temperature in an overhead stirred four liter polyethylene beaker. Normally, two liters of solution were tested. The speed of the stirrer, normally about 150 rpm, is just enough to suspend the solids. If aeration was needed, two fretted air diffusion tubes were immersed into the reactor. The amount of air introduced was about 4 l/min. The pH and the Oxidation/Reduction Potential (ORP) of the solution were periodically measured and recorded. After each specified time period, about 25 ml of solution was sampled and filtered through 0.45 micron filter paper. At the end of the test, the precipitates were either taken for settling tests or filtered. Sometimes, the softening test was performed after the filtration.

The settling test was performed in a two-liter graduated cylinder. The experiment and resulting calculations followed the modified Kynch method (5). All the settling tests in this research were performed without using a settling aid.

The Berkeley Pit water taken in October 1992 (see Table 1) was used for this bench scale test. The Pit water sample was taken from 200 feet below the surface (deep water) and 3 feet below the surface (shallow water). The sampled water was put into five-gallon plastic containers and was refrigerated at about 4°C before use. The water is a light greenish color.

Most of the chemicals were purchased from either Fisher Scientific or VWR Scientific corporations. Scrap iron, which was used for copper cementation, was supplied by Montana Resources (MR) which is currently mining the Continental pit.

Most of the metal concentrations were analyzed by ICP (Perkin-Elmer ICP/5500) in the Metallurgical Engineering Department at Montana Tech. Critical samples were analyzed by ICP (Perkin-Elmer Elan 5000 ICP-MS) in the Montana Bureau of Mines and Geology. The sulfate concentration was analyzed by the gravitational method using barium.

Neutralization and Precipitation

Acid neutralization is the most commonly used method to remove the acid, soluble heavy metals and other contaminants to acceptable limits of the water. Commonly used alkaline reagents to treat acid mine drainage are lime (CaO), limestone ($CaCO_3$), caustic soda ($NaOH$), soda ash (Na_2CO_3) and magnesium hydroxide ($Mg(OH)_2$). Because of the low cost and relatively easy operation, lime and limestone are the most widely used alkaline reagents. Since the solubility of Fe(III) is less than Fe(II), and ferric hydroxide has better settling characteristics, the neutralization processes are usually equipped with aeration to oxidize ferrous ions to ferric ions. Most of the metal ions, Al, Cu, Zn, Mn Cd and As, are either precipitated as hydroxides or co-precipitated with ferric hydroxide.

Chemistry of the Process. When lime and/or limestone reacts with acid mine drainage (AMD), the acid is neutralized. The heavy metals are removed from the solution by precipitation as metal hydroxides; sulfate can also be removed partially as gypsum. The overall reactions for these chemical treatment processes are shown as follows:

1. Acid neutralization with lime (CaO):

$$CaO + 2H^+ = Ca^{2+} + H_2O$$

2. Acid neutralization with limestone ($CaCO_3$) at pH < 6

$$CaCO_3 + 2H^+ = Ca^{2+} + H_2CO_3$$

3. Sulfate precipitation with lime or limestone

$$CaCO_3 + H_2SO_4 + 2H_2O = CaSO4 \cdot 2H_2O + H_2CO_3$$

$$CaO + H_2SO_4 + H_2O = CaSO_4 \cdot 2H_2O$$

4. Metal ion precipitation as metal ion hydroxides

$$M^{2+} + 2H_2O = M(OH)_{2(s)} + 2H^+$$

5. Oxidation of Iron (II) and precipitation of Fe(III)

$$4Fe^{2+}_{(aq)} + O_{2(g)} + 10\ H_2O\ =\ 4Fe(OH)_{3(s)} + 8H^+_{(aq)}$$

6. Adsorption and co-precipitation of metal ions with $Fe(OH)_{3(s)}$

$$Cu^{2+} + Fe(OH)_{3(s)} + H_2O\ =\ Fe(OH)_{3(s)} \cdot CuO_{(ad)} + 2H^+$$

Solubilities of Metal Hydroxides. Many metal ions undergo hydrolysis to form solid precipitates. The general reaction for metal hydroxide precipitation may be represented by the following formula:

$$M^{2+} + 2OH^-\ =\ M(OH)_2$$

The equilibrium concentration of the metal ions, and the amount of solid precipitate is governed by the equilibrium constant $\{OH^-\}^2\{M^{2+}\}$, which is equal to the inverse of the solubility product, Ksp, reported in the literature *(6)*. Since simple M^{2+} forms various complexes with hydroxide ions and anions, such as sulfates, the concentration of M^{2+} is controlled by all the equilibrium and mass balance equations involved. The sum of the total concentrations of the metal ions, including complexes that equilibrated with the solid, is defined as the solubility of that solid. Figure 1 shows the solubilities of major metals present in the Berkeley Pit water, which were calculated using the STABCAL program *(7)*.

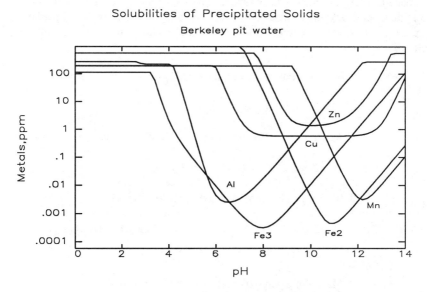

Figure 1. Solubilities vs. pH of Various Metals Present in the
Berkeley Pit Water.

Adsorption of Metals by Ferric Hydroxide. Iron and aluminum hydroxides are known and well documented for adsorbing trace metals from water. Ferric hydroxide in particular is capable of adsorbing both cations such as Cd^{2+} and anions such as AsO_4^{3-}. It has a tremendous amount of surface area, 600 m2 per gram of iron hydroxide, and high adsorption capacity, 0.2 moles of metal compound per mole of ferric hydroxide [8][9]. The success of removing one particular metal depends on the amount of the adsorbent available, the pH of the water, and the competitiveness of all the ions involved in the system.

The adsorption reactions on ferric hydroxide, along with hydroxides precipitation for the Berkeley Pit water as a function of pH, were simulated by using the STABCAL program. The calculations were based on the chemical composition of the water taken in 1992 (see Table 1). Species involved (solid, aqueous and adsorbed by ferric hydroxide) and adsorption capacity of ferric hydroxide are identical to those in Reference [2].

The residual concentrations of As, Cd and Pb from computer simulation due to precipitation with and without adsorption were plotted in Figure 2. The calculation assumed that 95% of ferrous ions are oxidized to ferric ions. The symbol (a) indicates where adsorption is considered. The results show that arsenic cannot be removed by precipitation but only by adsorption. Adsorption removes lead and cadmium at lower pH levels.

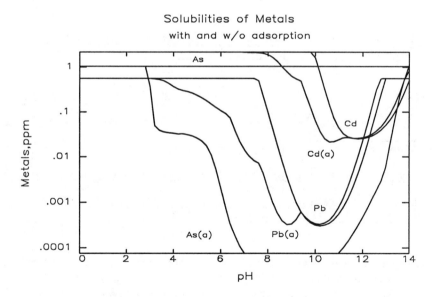

Figure 2. Concentrations of As, Cd and Pb vs. pH with and without Considering the Adsorption onto Iron Hydroxide Surface

One-Stage Neutralization

The purpose of the one-stage neutralization was to demonstrate metal removal by using the conventional neutralization process with various types of alkaline reagents with or without aeration. The reagents used in these tests included lime, limestone and magnesium hydroxide.

Neutralization Without Aeration. One-stage neutralization without aeration was investigated using three different reagents: lime, caustic soda and magnesium hydroxide. The conditions and results from critical tests are summarized in Table 2. The results indicate that none of the reagents can produce water clean enough to meet EPA Gold Book criteria. Magnesium hydroxide produces the fastest settling and most filterable sludge, followed by lime. Caustic soda produces the slowest settling and filterable sludge which entrapped the most water. Magnesium hydroxide, however, is unable to raise the pH of the water greater than 9.

Neutralization With Aeration. The one-stage neutralization with aeration testing involved only lime and magnesium hydroxide. The conditions and results from critical tests are also summarized in Table 2. Compared with no aeration, aeration improves the removal of the metals in the water and produces faster settling and smaller volumes of sludge. The results, however, still show that one-stage neutralization with aeration is still incapable of cleaning the water enough to pass EPA Gold Book Criteria.

Table 2. Results from Critical Tests of One-Stage Neutralization (ppm)

Reagent (g/2L)	pH	[As]	[Ni]	[Cu]	[Zn]	[Mn]	[Al]
NON-AERATION							
Lime (9.7)	10.32	0.005	0.048	0.018	0.097	0.012	1.163
NaOH (9.5)	8.79	0.006	0.008	0.413	1.094	7.012	1.020
Mg(OH)$_2$ (26)	7.39	0.001	0.412	0.022	37.720	147.8	<0.02
AERATION							
Lime(7.5)	9.08	<0.001	0.034	0.006	0.061	0.958	0.555
Lime(9)	10.0	0.001	0.028	0.005	0.087	0.065	3.013
Lime(9)	10.13	<0.001	0.024	0.010	0.048	0.005	3.900
Lime(10.4)	11.14	0.002	0.026	0.032	0.262	0.017	0.318
Mg(OH)$_2$ (30)	8.10	0.002	0.056	0.057	1.750	89.480	0.350

Problems Related to One-Stage Neutralization. One-stage neutralization is a simple, effective and proven technique to treat acid mine drainage, particularly from coal mines. The problem of using this method to treat the Berkeley Pit water is to identify the pH that will produce water that satisfy Mn, Zn and Cd regulations while meeting the Al regulation level. The solubilities of Mn and Al hydroxides as a function of pH considering all the components in the Berkeley Pit water are computed and shown in Figure 3. Judging from the diagram, the best pH to remove Al is between 6 and 7, which is too low to remove Mn. When pH is raised high enough to remove Mn (greater than 10), Al precipitate starts to redissolve.

There were several options considered which have the potential to satisfy EPA water regulations for Mn and Al: the first concern is to separate the Al precipitates by controlling the pH to around 6, then raising the pH to over 10 to precipitate Mn; the second is to separate the precipitates from treated water by one-stage neutralization, then bring the pH back to about 6 in order to precipitate re dissolved Al; and finally, to develop a process which will have the capacity to reduce the precipitation pH for Mn. For example, improving the absorption effects by increasing the concentration of ferric ions in the solution.

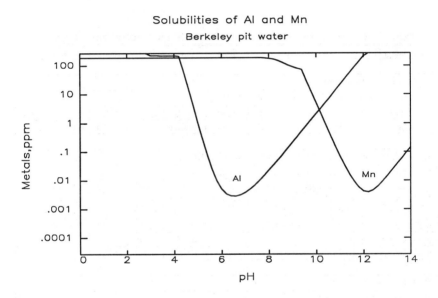

Figure 3. Solubilities of Al and Mn vs. pH

Two-Stage Neutralization

Metal hydroxides associated with the Berkeley Pit water precipitate at various pH levels. The two-stage neutralization method was therefore proposed and investigated. The purpose was to precipitate most of the Fe(III), Al and Cu at a pH of about 6. After separating the precipitates from the water, the pH was further raised to 10.25-10.5 to remove the rest of the metals including Zn, Mn and Cd.

Test Conditions. Different conditions and reagents were used for two-stage neutralization tests. These conditions were:

A. Aeration in both stages
    A1. First-stage (lime) and second-stage (lime or NaOH + Na_2CO_3)
    A2. First-stage (limestone) and second-stage (lime or NaOH + Na_2CO_3)
B. Non-Aeration in Both Stages
    B1. First-stage (limestone) and second-stage(lime)
C. Aeration only in one of the stages: lime for both stages
    C1. Aeration in the first-stage
    C2. Aeration in the second-stage

Results Summary. The ICP results from some critical tests are summarized in Table 3. Using the reagents and conditions stated, two-stage neutralization is capable of removing Fe, Al, Pb, As, Cd, Zn, Cu and Mn in the Berkeley Pit water to meet EPA Gold Book standards.

Table 3. Results from Critical Tests of Two-Stage Neutralization (ppm)

1st-stage pH	2nd-stage pH	[Al]	[Cd]	[Cu]	[Fe]	[Mn]	[Zn]
A. Aeration in Both Stage							
A1. 1st-Stage (lime) and 2nd-Stage (lime + Na_2CO_3)							
5.69 (lime 5.0)	10.37 (lime 3.2 SA 3.2)	0.098	<0.001	0.011	0.013	0.004	0.106
5.64 (lime 4.9)	10.16 (lime 3.0 SA 3.2)	0.094	<0.001	0.002	0.019	0.010	0.107
A2. 1st-Stage (limestone) and 2nd-Stage (lime or NaOH + Na_2CO_3)							
6.00 (LS 6.5)	9.89 (lime 2.6 SA 5.5)	<0.022	<0.002	0.022	0.015	0.085	0.085
6.08 (LS 6.5)	8.92 (NaOH 0.4 SA 5.5)	<0.022	0.013	0.047	0.053	27.64	0.455
6.11 (LS 6.5)	9.89 (NaOH 2.9 SA 5.5)	<0.022	<0.002	0.152	0.028	0.089	0.860
6.06 (LS 6.5)	8.98 (NaOH 1.4 SA 2.4)	0.095	0.052	0.373	1.54	22.16	5.984
6.15 (LS 6.5)	9.97 (NaOH 3.6 SA 2.4)	<0.022	<0.002	0.054	0.032	0.043	0.124
6.06 (LS 6.5)	10.38 (NaOH 5.0)	0.022	<0.001	0.019	<0.04	0.003	0.081
6.06 (LS 6.5)	10.26 (NaOH 4.8)	0.054	<0.001	0.011	<0.04	0.012	0.094
B. Aeration only in One-Stage							
B1. Aeration in 1st-Stage							
5.49 (lime 4.2)	10.51 (lime 2.84)	0.128	<0.001	0.020	0.030	0.365	0.072
5.49 (lime 4.2)	10.25 (lime 2.46)	0.107	<0.001	0.025	0.111	0.733	0.086
B2. Aeration in 2nd-Stage							
5.89 (lime 5.2)	10.40 (lime 4.93)	0.106	<0.001	0.036	0.029	0.047	0.140
5.89 (lime 5.2)	11.66 (lime 6.53)	0.132	<0.001	0.009	0.018	<0.01	0.097
5.75 (lime 4.3)	10.49 (lime 3.90)	0.097	<0.001	0.019	0.052	<0.01	0.118
5.91 (lime 4.3)	10.54 (lime 4.26)	0.097	<0.001	0.073	0.107	0.027	0.107
5.83 (lime 4.2)	10.26 (lime 3.9)	0.025	<0.001	0.011	<0.04	0.006	0.077
5.83 (lime 4.2)	10.16(lime 4.1)	0.032	<0.001	0.025	<0.04	<0.01	0.099

The pH columns also indicate the reagent and amount used based on two liters of water. Concentrations of Pb < 3 ppb, As < 2 ppb and Ag < 1 ppb. LS stands for limestone, SA for soda ash.

The pH of the first-stage should be kept around 6. Lime and limestone are both effective. Limestone, however, is probably the better choice since it controls and buffers the pH of the Berkeley Pit water to around 6. In addition, limestone produces faster settling and more filterable sludge. A small amount of lime can be added along with limestone if limestone alone cannot raise the pH high enough.

Aeration is necessary during the second-stage of neutralization. Both lime and caustic soda are effective for the second-stage. The pH of the second-stage should be between 10 and 10.5, and is best around 10.25. It is also acceptable to have a pH greater than 10.5 if the pH of the first-stage is slightly higher than 6 to remove as much Al as possible.

Second-stage neutralization produces a relatively large volume of sludge. For two liters of the Berkeley Pit water, the second-stage generates about 500-600 ml (11-12 grams of dried weight) of sludge after 24 hours of settling. The first-stage created only 100-150 ml (7.5-8.5 grams of dried weight). Without flocculant, sludge from the second-stage settles slowly. Using the results from settling tests, the modified Kynch method estimated that the sludge needs a thickening area about 2,000 cm^2 per gram of dry sludge per minute.

Large Scale Two-Stage Neutralization Test

The two-stage neutralization method was demonstrated and verified by testing 80 gallons of the Berkeley Pit water. The water was divided into two batches, 40 gallons each, and treated in a 60 gallon plastic container with an air driven stirrer. The conditions and measured values for each test are listed in Table 4. The ICP results from critical samples are listed in Table 5.

Table 4. Conditions of Two-Stage Neutralization of an 80 gallon Water (ppm)

Before the Treatment				After the Treatment		
Test	gallon	pH	Eh(mv)	Reagent	pH	Eh(mv)
A. 1st-Stage Neutralization						
Batch 1	40	2.92	670	912g limestone and 70g lime	6.05	439
Batch 2	40	2.93	650	912g limestone and 70g lime	6.07	430
B. 2nd-Stage Neutralization						
Batch 1	40	5.88	496	285g lime	10.39	396
Batch 2	37.5			260g lime	10.37	394
C. Softening						
Batch 1	30			238g soda ash	9.25	495
Batch 2	26.5			210g soda ash	9.38	500
D. Acidification						
Batch 1	55	9.25		1 ml concentrate sulfuric acid	7.43	

Table 5. Results from Two-Stage Neutralization of an 80 gallon Water (ppm)

Sample #	pH	[Al]	[Mn]	[Cu]	[Zn]	[As]	[Cd]	[Pb]
Neutralization	10.37	0.070	<0.001	<0.001	0.080	<0.001	<0.001	<0.001
Softening	9.25	0.047	<0.001	0.006	0.059	<0.001	<0.001	<0.001
Acidification	7.43	0.006	<0.001	0.016	0.033	<0.001	<0.001	<0.001
Polishing		<0.05	<0.005	<0.010	<0.010	<0.005	<0.003	<0.002

First-Stage Neutralization. Each batch of 40 gallons of water was neutralized with 912 grams of limestone followed by 70 grams of lime, and was run for two hours without aeration. After settling for about two hours, each batch of water was pressure filtered with Whatman #42 filter papers. Total weight of the filter cake including the moisture was 17.7 kilograms. The filtrate sat overnight.

Second-Stage Neutralization. The test was carried out again in the same 60 gallon container. In order to aerate the solution, the pressurized air was introduced through four glass tubes. Two batch tests were performed (see Table 4). Each batch contained about half the volume of solution that was collected from each batch during the first-stage treatment. Lime was used to adjust the pH of the solution to about 10.25.

After two hours of reaction and one day of settling, the solution was pressure filtered with Whatman #42 paper. The solution was again split into two containers. Total weight of the filter cake, including moisture, was 23.7 kilograms. The liquid sample was taken, analyzed and listed in Table 5. The alkalinity was only 25.57 ppm for $CaCO_3$.

Softening. The softening test was performed in the 60 gallon container. Again the tests were carried out in two batches (see Table 4). After adding the soda ash reagent, the solution was left overnight then pressure filtered. Total weight of the filter cake including the moisture was 520 grams. The results from the softening test are listed also in Table 5.

Acidification. The final solution was acidified with 1 ml of concentrated sulfuric acid. The pH of the solution was reduced from 9.25 to 7.43. The results from acidification are listed also in Table 5. Based on the Ca and Mg concentrations, the hardness of the solution after softening and acidification was estimated to be about 220 ppm.

Polishing step. Solution, after acidification, was sent to Osmonics Inc. in Minnesota for final polishing tests to remove residual sulfate and Total Dissolved Solids (TDS). A microfilter, OSMO-19T60-SS97C-PES, was used with nanofilter, SX10 NF. Under the condition of 85% water recovery, sulfate was reduced from 2200 to 270 ppm and TDS was reduced from 3880 ppm to 520 ppm. Other metal concentrations in the permeate are listed on Table 5.

Copper Cementation Prior to Neutralization

Among the metals present in the Berkeley Pit water, copper (over 200 ppm) is probably the most valuable one. The purpose of this test was first to investigate the possibility of using industrial scrap iron to produce cement copper, and second to identify any ill effects from cementation on the subsequent neutralization step. The results have been used to determine the effectiveness of the treatment processes and to calculate the iron consumption.

The cementation operation was and still is practiced near the Berkeley Pit by taking the leach water that percolated through the waste rock dumps from mining the Berkeley Pit. Montana Resources currently mines the Continental pit and mills in the Weed Concentrator. It also operates the leach-precipitation plant. The Berkeley Pit water, in fact, contains higher copper concentration than the leach solution.

Copper Cementation with Scrap Iron. Scrap iron, provided by MR, was first cut to one inch square, then soaked in 10% HCl for two days to remove the plastic film on the surface. The cementation test was carried out by using two liters of solution in a four liter overhead beaker, stirred at 650 rpm; the beaker contained 30 grams of scrap iron. About 25 ml of the solution was taken and filtered at 30, 60 and 90 minutes for analyses. Results from using the Berkeley Pit water as well as the leach solution are listed in Table 6. One hour of retention time is capable of removing over 90% of the copper from both the Berkeley Pit water and the leach solution.

Neutralization after Copper Cementation. The effectiveness of neutralization after copper cementation with scrap iron was examined. The solution was tested using the one-stage neutralization method with lime addition for two hours. Results, shown in Table 7, are quite comparable to the solution without cementation in terms of the amount of lime used, final pH, and concentrations of residual metals.

Table 6. Results from Copper Cementation for Various Sources of Water (ppm)

Source of water	pH	Eh(mv)	[Fe]	[Cu]	% [Cu] removed
A. Berkeley Pit Water	3.08	660	1030	215.7	
0.5 Hour	3.91	520	1309	65.03	69.85
1.0 Hour	4.45	440	1421	15.34	92.89
1.5 Hour	4.45	400	1441	11.30	94.76
B. Shallow Pit Water	2.85	773	279	189.2	
0.5 Hour	4.09	538	567	22.46	88.13
1.0 Hour	4.40	461	642	4.55	97.60
1.5 Hour	4.41	368	707	3.44	98.18
C. Leach Solution	2.76	643	1126	103.1	
0.5 Hour	3.54	520	1599	18.83	81.73
1.0 Hour	3.85	370	1769	3.18	96.92
1.5 Hour	3.96	326	1884	0.80	99.22

Table 7. Comparison with and without Copper Cementation (ppm)

Test condition	lime g/2L	pH	Eh (mv)	[Cu]	[Fe]	[Mn]	[Al]
With Cementation	9.0	9.8	433	0.04	<0.04	<0.04	<0.1
W/O Cementation	9.0	10.00	366	<0.03	<0.04	<0.04	0.27

SUMMARY

Since 1982, 5 to 7 million gallons per day of the acid mine water has flowed into the Berkeley Pit. The water is very acidic and contains heavy metals that are several orders of magnitude greater than EPA discharge regulations. The water in the Berkeley Pit will create a serious threat to the Columbia river drainage, the ground water quality, and the public health and well-being of the people in the Butte, Montana area. From the hygienical and esthetical points of view, the water needs to be treated and cleaned up in the near future.

Chemistry of the water

1. The Berkeley Pit water is very acidic (pH 2.5-3.1), has a high oxidation potential (Eh 600-800 mv), and is highly saturated with Fe, Al, Si, Ca, K, and sulfate. Precipitation is occurring.
2. Sulfate is the main anion in the water. An average of 50% of all metals are complexed with sulfate. The concentrations of arsenic and lead are relatively low, especially from the water near the surface.
3. Iron is the main cation in the water. Most of the iron exists as ferrous ions, except near the surface, which is easily oxidized. Other metal ions may be catalyzing the oxidation reaction. The oxidation precipitates some of the iron and produces extra acid in the water.

Neutralization Treatment

1. The Berkeley Pit water can be treated by neutralization and chemical precipitation processes. Most of the metals and contaminants can be removed from the solution. The multiple-stage operation is necessary in order for the water to meet the present Gold Book standards.
2. The water can be treated using the two-stage neutralization process. The first-stage uses limestone, with extra lime if necessary, to neutralize the water to a pH of about 6 to precipitate most of the Al and part of the Fe and Cu. After separating the solid, the pH of the solution is further raised, with aeration, to 10.25 to precipitate the rest of the metals in the water.
3. The two-stage neutralization processes were tested and verified by using 80 gallons of the Berkeley Pit water.
4. Copper present in the water can be cemented out with scrap iron. After copper cementation, the solution can be treated with neutralization processes. No ill effects due to the cementation have been identified.

ACKNOWLEDGMENTS

The authors would like to thank:

USBM Generic Center Research (#G1175149-3022) and Bill Williams, Sandy Stash and Dave Sinkbeil (ARCO Superfund) for financial support.

Anne Lewis-Russ (Canonie Environmental), Russ Forba and Mike Bishop (EPA Montana Operations), Ray Tilman (Montana Resources), James Scott (Montana Department of Health and Environmental Sciences), Sam Worcester (Metallurgical Engineering Department, Montana Tech), Ted Duaime (Montana Bureau of Mines and Geology) and Ray Tilman (Montana Resource) for technical support.

Steve McGrath (Montana Bureau of Mines and Geology) and Haiyang Gu and Jin Wang (Metallurgical Engineering Department, Montana Tech) for chemical analysis.

REFERENCES

1. Qi Liu and H.H. Huang, *Metallurgical Processes for the Early Twenty-First Century;* Editor, H.Y.Sohn; Technology Practice, 1994, Vol. I; pp 695-713.
2. A. O. Davies (Camp Dresser & McKee), *Factors Affecting the Geochemistry of the Berkeley Pit, Butte, Montana,* EPA Work Assignment No.: 373-8L22, U.S. EPA Contract No.: 68-01-6939, REM II Document No. 292-PP1-EP-FUNK (May 1988).
3. Qi Liu, Thesis: *Bench Scale Chemical Treatability Study of the Berkeley Pit Water in Mine Waste Technology Pilot Program;* Montana Tech, Butte, MT, 1994.
4. Canonie Environmental, *Treatability Sampling and Bench-Scale Testing Report: Butte Mine Flooding Operable Unit, Butte Montana,* Prepared for ARCO, 1993.
5. B. A. Wills, *Dewatering in Mineral Processing Technology,* 4th Ed.; Pergamon Press; Oxford, England, 1988.
6. Werner Stumm & James J. Morgan, *Aquatic Chemistry;* 2nd Ed.; John Wiley & Sons, Inc.; New York, 1981; pp 238-249.
7. H. H. Huang, *STABCAL - Stability Calculation for Aqueous Systems,* Montana Tech, Butte, MT, 1993.
8. J. D. Allison, et. al., *MINTEQA2/PRODEFA2, A Geochemical Assessment Model for Environmental Systems,* Ver. 3, Environmental Research Laboratory, EPA, 1990.
9. D. A. Dzombak and F. M. Morel, *Surface Complexation Modeling Hydrous Ferric Oxide;* John Wiley & Sons, Inc., New York, 1990; pp 73-78.

RECEIVED March 14, 1995

Chapter 17

Chelating Extraction of Zinc from Soil Using N-(2-Acetamido)iminodiacetic Acid

Andrew P. Hong, Ting-Chien Chen, and Robert W. Okey

Department of Civil Engineering, University of Utah,
Salt Lake City, UT 84112

N-(2-acetamido)iminodiacetic acid (ADA) was used to extract zinc from a spiked soil and release the metal subsequently. It was hypothesized and tested experimentally that through a proper choice of chelator: 1) extraction could be made selective toward heavy metals; and 2) extracted metals could be readily recovered as oxide, hydroxide, and/or carbonate precipitates by increasing the solution pH. The results have shown that 1) ADA is able to extract zinc from a spiked soil and subsequently release it upon favorable conditions; 2) ADA selectively removes zinc on the presence of competing Fe and Ca ions. Chemical equilibrium modeling was used as a predictive tool and was found to be useful for selecting suitable chelators and treatment conditions for the extraction and recovery of metals.

Heavy metal contamination of soil is a common problem encountered at many hazardous waste sites. Lead, chromium, cadmium, copper, zinc, and mercury are among the most commonly observed contaminants. Once released into the soil matrix, heavy metals are often strongly retained, making remediation difficult. Various techniques have been used for remediation of contaminated soils (1-3). The use of chelators, in-situ or during soil washing following excavation, to extract heavy metals from contaminated soils is seen as a treatment method. Ethylenediaminetetraacetic acid (EDTA) was most often researched (4-8). Nitrilotriacetic acid (NTA) (6,9), diethylenediaminepentaacetic acid (DTPA), ethyleneglycolbis(ethylamine)tetraacetic acid (EGTA) (10), hydroxylamine (7), and citrate (11) were also considered for mobilizing heavy metals in soils. Although sorption studies of metals to mineral solid phases in the presence of synthetic or naturally occurring chelates have been conducted over the past two decades (12-16), the application of chelators in extracting heavy metals from contaminated soils as a

0097–6156/95/0607–0210$12.00/0

treatment has been relatively recent. Only a few chelators have been tried for this application.

A potentially useful chelator must be powerful enough to overcome the competition for metals due to soil surface adsorption and/or surface precipitation, yet the resulting mobile complex should be readily treated to release the metal as to facilitate reuse of the chelator.   Furthermore, a suitable chelator should prefer complexing with heavy metals of interest over ambient competing cations.   For this study, we have selected ADA as the chelator of choice because 1) it has high metal complexing abilities; 2) it contains a nitrogen ligand atom that prefers complexing with soft sphere (B-type) cations including Zn^{II}; and 3) it has multiple coordinating sites (ligand atoms) that typically form very stable complexes.   We report here the extraction of zinc from a laboratory spiked soil using chelator N-(2-acetamido)iminodiacetic acid (ADA).   The extracted zinc was subsequently recovered as a zinc oxide precipitate upon raising the solution pH, while the chelator remained soluble in the aqueous phase.   Chemical equilibrium modeling that has led to our selection of ADA for zinc extraction and recovery is presented.

Methods and Materials

Deionized water (18 MΩ-cm) was used in all procedures and was obtained from a Milli-Q system (Millipore).   The chelator ADA (Fluka) was used as received.   Soils were taken from a Salt Lake City site, air-dried for one month, then passed through a 2-mm sieve.   Some properties of the soil were characterized and shown in Table I.   Typical experiments were conducted in 125 mL glass Erlenmeyer flasks using a batch solution volume (V) of 100 mL.   No difference in results was seen between experiments conducted using glass flasks and polyethylene flasks.   Ionic strength (I) was held constant at 0.1 M using $NaClO_4$ (99%); total carbonate concentration (C_T) was added at 1 mM using $NaHCO_3$ unless stated otherwise.   All flasks were sealed with stoppers to reduce CO_2 exchange with the atmosphere during experiment.   All pH adjustments were performed manually with either a 5 M HNO_3 or NaOH solution, and pH measurements with an Orion model SA 720 pH meter.   A 1000 mg/L Zn stock solution was following ASTM method D1691.   A gyratory shake table (New Brunswick Scientific Co., Model G-2) was used to provide agitation during adsorption and extraction procedures.   The soil was kept in suspension by operating the shake table at 260 rpm.   All experiments were conducted at the room temperature of 23±1 °C.   In measuring aqueous metal concentrations, aliquots were withdrawn from the reaction mixtures, filtered through a 0.45 μm filter (Gelman Sciences sterile acrodisc), and acidified with nitric acid.   Zn concentration was analyzed by atomic absorption (AA) spectrometry (Perkin Elmer Model 280) using ASTM method D1691.   Standard procedures were followed when available (*22, 23*).   All equilibrium calculations were performed using constants as listed in Tables II.

Typical concentration conditions in adsorption and extraction experiments were 50 mg/L Zn, I = 0.1 M, C_T = 1 mM, and V = 100 mL.   A known amount of soil was added to a solution of desirable Me, C_T, I, and pH conditions.   The pH was initially adjusted if necessary and the mixture was continuously agitated and maintained in suspension by a shake table.   Initial pH values (pH_0) between 4.5 and

Table I. Characteristics of Soil

Parameter	Value		Procedure
Soil pH	7.8 in water		(17)
	7.5 in CaCl$_2$		(17)
pH$_{zpc}$	8.1		(18)
Specific gravity	2.65		(19)
Composition (<2mm)			(19)
	90.4% sand		
	7.5% silt		
	2.1% clay		
Cation exchange capacity	13.9 meq/100 g		(20)
Ambient metal (mg/Kg soil)	Fe	Zn	(21)
exchangeable	0.6	0.6	
adsorbed	2.75	1.1	
organic	73.8	35.8	
carbonate	40.1	107.1	
sulfide/residual	5617.5	271.3	
total	5735	415.9	

Table II. Equilibrium Reactions of Zinc in a Carbonate-Bearing Water (*24*)

Equilibrium Reaction	log K, 25°C, I=0
$ZnCO_3(s) = Zn^{2+} + CO_3^{2-}$	-10.0
$Zn(OH)_2(s) = Zn^{2+} + 2OH^-$	-15.52 to -16.46
$ZnO(s) + H_2O = Zn^{2+} + 2OH^-$	-16.66
$Zn^{2+} + CO_3^{2-} = ZnCO_3^°(aq)$	5.11
$Zn^{2+} + HCO_3^- = ZnHCO_3^+$	0.7
$Zn^{2+} + OH^- = ZnOH^+$	5.0
$Zn^{2+} + 2OH^- = Zn(OH)_2^°(aq)$	11.1
$Zn^{2+} + 3OH^- = Zn(OH)_3^-$	13.9
$Zn^{2+} + 4OH^- = Zn(OH)_4^{2-}$	15.5
$2Zn^{2+} + OH^- = Zn_2OH^{3+}$	5.5[a]
$4Zn^{2+} + 4OH^- = Zn_4(OH)_4^{4+}$	27.9[a]
$H^+ + OH^- = H_2O$	14.0
$H^+ + CO_3^{2-} = HCO_3^-$	10.33
$H^+ + HCO_3^- = H_2CO_3^°$	6.35
$H^+ + ADA^{2-} = HADA^-$	6.84
$H^+ + HADA^- = H_2ADA^°$	2.31[b]
$Zn^{2+} + ADA^{2-} = ZnADA^°$	7.10[b]
$Zn^{2+} + 2ADA^{2-} = Zn(ADA)_2^{2-}$	9.22[b]
$Zn(H_{-1}ADA)ADA^{3-} + H^+ = Zn(ADA)_2^{2-}$	9.49[b]
$Zn(H_{-1}ADA)_2^{4-} + H^+ = Zn(H_{-1}ADA)ADA^{3-}$	10.56[b]

[a] 25°C, I=3.0;   [b] 25°C, I=1.0

6.1 were used to ensure that 50 mg/L Zn was completely dissolved and that the soil was introduced to a homogeneous metal solution. Adsorption and extraction tests were performed consecutively by adding the desired amount of chelator (L_T) to the suspension following the termination of the adsorption run. An equilibration time of 24 h was allowed for batch adsorption and extraction runs prior to metal analysis.

Extraction of Zn(II) was performed in the presence of excess Fe(III) and Ca(II) competing ions. The adsorption and extraction procedures followed those described above, but in spiking the soil, $CaCl_2 \cdot 2H_2O$ and $Fe(NO_3)_3 \cdot 9H_2O$ were added in various mole ratios to Zn being studied. Other concentration conditions were: Zn_{added} = 50 mg/L, soil = 20 g/L, I = 0.1 M, C_T = 1 mM, V = 100 mL, and pH = 4.9-5.2.

The pH of the Zn-ADA complex solution was raised to recover the metal as ZnO, as identified by X-ray diffraction measurement. The conditions were: C_T = 1 mM, I = 0.1 M, Ca = 10 mM, V = 100 mL, and ADA:Zn mole ratio = 6.5:1. A 12-h period was allowed for precipitation.

Results and Discussion

Extraction of Zinc from Soil. The extraction of zinc from spiked soils using ADA was studied by measuring the extracted aqueous phase Zn (i.e., total dissolved zinc, Zn_T) under different pH, soil suspension, total chelator concentration (L_T), total carbonate concentration (C_T), and age of the spiked soil conditions.

Dependence on pH and Soil Suspension. For initial pH between 4.4 and 6.1, Figure 1 shows the adsorption of zinc to soil and the subsequent desorption of zinc from soil when chelate ADA is added. As indicated, between 1,500 ppm and 4,000 ppm of zinc was loaded onto the soil under various soil suspension and initial pH conditions. The experiments were carried out at C_T = 1 mM and I = 0.1 M but with varying amounts of suspended soil (0.03 to 2.4 g) in 100 mL of deionized water. The total zinc used (Zn_{added}) was 50 mg/L and the ADA added was 1 mM, corresponding to a mole ratio of 1.3 for L_T/Zn_{added}.

Dependence on L_T. Figure 2 shows the extent of extraction of zinc from soil as a function of L_T/Zn_{added}. Under the specified conditions, 2 mM of ADA which corresponds to a L_T/Zn_{added} mole ratio of 2.6 was able to extract, thereby solubilize, 97% of the total zinc from the spiked soil.

Dependence on C_T. The adsorption and extraction behavior were studied in the presence of various amounts of carbonate species; the results were not influenced by C_T in the range 0 to 10 mM.

Effect of Age of Spiked Soil and Stability of Soluble Complex. Figure 3 summarizes results of a series of experiments undertaken over a 30-day period to 1) determine whether the spiked soil age exerted an effect on the extraction efficiency, and to 2) assess the stability of the concomitantly formed soluble complex on prolonged aerobic contact with soil. Four 200 mL batches of 50 mg/L Zn_{added} solution at pH 6.0 were added with 2 g of soil each (Day 1). After 24 h contact

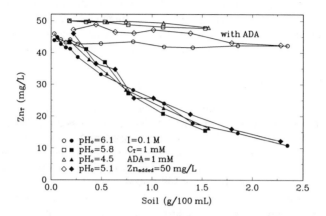

Figure 1   Adsorption (solid symbols) of zinc to soil and desorption (open symbols) upon addition of ADA.

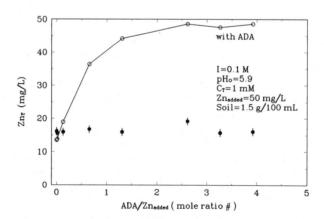

Figure 2   Adsorption (solid symbols) of zinc to soil and extraction (open symbols) upon addition of ADA as a function of L_T/Zn_{added} mole ratio.

with the suspended soil, 60% of the Zn became associated with the soil matrix (Day 2). A powder ADA was then added at 2 mM to one batch at a time at Days 2, 10, 14, and 29. The Zn_T of all batches before, during, and after the ADA addition were monitored and shown in Figure 3. It showed that the extraction efficiency did not depend on the age of the spiked soil over a one-month period, and that the soluble complex remained stable over the same period. However, it should be cautioned that the extraction efficiency should probably be studied over a longer period.

Recovery of Metal and Chelator

Release of Zn from ADA. Figure 4 shows results of experiments conducted to test whether zinc can be released from ADA and be separated as a precipitate at higher pH values. Up to pH 9.5 as shown, 8.5 mg/L (0.13 mM) of zinc was completely solubilized by 1 mM ADA. When the pH of the Zn-ADA complex solution was increased to 10, a precipitate appeared and the Zn_T significantly decreased. The precipitation appeared to be most extensive at pH 11, above which pH resolubilization appeared to occur. The addition of Ca was necessary to aid the separation of Zn from ADA; without the presence of Ca, Zn could not be recovered from ADA even at pH as high as 12.7. The progress of precipitation was followed by measuring the decreasing Zn_T as a function of time. In most cases, 90% of the chelated Zn could be recovered as precipitate within 2 h. The collected solid precipitate was analyzed by X-ray diffraction method and was found to be ZnO(s).

Repeated Use of Reclaimed Chelator. Consecutive extraction experiments were conducted using reclaimed ADA from preceding runs. Four batches of soil suspension (2 g/100 mL each) were individually spiked with about 10 mg/L Zn_{added}. The spiked soils were then centrifuged and collected. The first batch was extracted for 24 h with 100 mL of 1 mM ADA solution and then the Zn_T measured. The extracted soil was separated by centrifugation, then 20 mM $CaCl_2 \cdot 2H_2O$ was added to the resulting complex solution. The solution pH was adjusted to 11.5 and precipitation was allowed to occur for 24 hours. After this time Zn_T was measured and the formed precipitate was separated; the clear chelator solution was adjusted to pH 6 and then applied to the second batch of the Zn loaded soil. The entire extraction and separation procedure was repeated for three more times. The extraction results of ADA toward Zn as the chelator was repeatedly applied to fresh batches of Zn loaded soil showed the extraction efficiencies (mg Zn extracted/mg Zn initially adsorbed) of the consecutive runs to be 100, 71, 72, and 79%.

Selectivity Toward Zinc. The extraction preference of ADA toward Zn^{II} over Ca and Fe^{III} was assessed by performing extraction experiments in the presence of added Ca and Fe^{III} at various mole ratios to Zn, i.e., mole ratios of Ca_{added}/Zn_{added} and Fe_{added}/Zn_{added}. ADA was added to batches of soil spiked with excess $CaCl_2 \cdot 2H_2O$ or $Fe(NO_3)_3 \cdot 9H_2O$ (at 1 to 10 folds of Zn_{added}) with pH adjusted to about 5 initially (pH subsequently increased to between 6.9 and 7.3 in the presence of soil). The results of Figure 5 show that the presence of Ca (up to a Ca/Zn mole ratio of 10) reduces soil's uptake of Zn but has no effect on ADA's extraction of Zn from soil. On the contrary, a 10-fold mole excess of Fe over Zn in the system appears to have

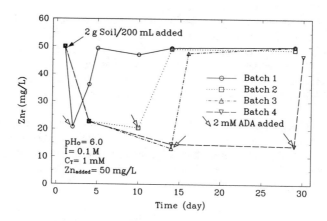

Figure 3   Total aqueous zinc concentration observed before, during, and after ADA additions over a 30-day period, illustrating i) negligible effect due to the age of the spiked soil and ii) the stability of the soluble complex.

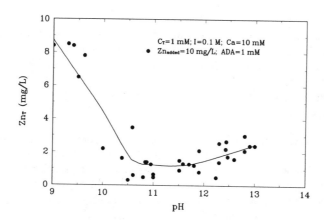

Figure 4   Release of complexed Zn from ADA at various pH.

no effect on Zn uptake by soil but to reduce the extraction of zinc by ADA moderately. It should be noted that the mole ratios under the conditions studied were Zn:ADA:Ca (or Fe) = 1:2.6:1 to 10.

We have demonstrated the ability of ADA to extract Zn from soil and subsequently to release the metal as a solid precipitate. The selection of ADA has been based on its high complexing ability and, more importantly, its ability to release the extracted metal reversibly. Other factors to be considered for chelating extraction includes 1) equilibrium chemistry of the metal, 2) soil-metal interaction at the solid-liquid interface, and 3) metal-ligand complexation reactions.

Equilibrium Chemistry of Zinc. The equilibrium chemistry important for zinc in a carbonate-bearing natural water is summarized in Table II. When its aqueous solubility is exceeded, zinc can exist in solid forms as oxide, hydrous oxide, and/or carbonate. The log function of activity ratios for $ZnCO_3(s)$, $ZnO(s)$, and $Zn(OH)_2(s,\varepsilon)$ to Zn^{2+} can be defined as:

$$\log \frac{\{ZnCO_3(s)\}}{\{Zn^{2+}\}} = 10 + \log C_T + \log \alpha_2 \qquad (1)$$

$$\log \frac{\{Zn(OH)_2(s)\}}{\{Zn^{2+}\}} = 16.5 + 2 \log K_w + 2 \, pH \qquad (2)$$

$$\log \frac{\{ZnO(s)\}}{\{Zn^{2+}\}} = 16.7 + 2 \log K_w + 2 \, pH \qquad (3)$$

A comparison of the activity ratio curves plotted in Figure 6 suggests that below pH 9, $ZnCO_3(s)$ is a predominant solid controlling the solubility of zinc while above pH 9 $ZnO(s)$ is the controlling solid. Therefore, as the solubility is exceeded, zinc is predicted to precipitate as $ZnCO_3(s)$ at pH < 9 or as $ZnO(s)$ at pH > 9. This conclusion is confirmed by X-ray diffraction results for $ZnO(s)$ precipitate recovered at high pH (e.g., 12).

Soil-Metal Interaction. When soil is present, it can accommodate the aqueous and particulate forms of zinc. Particulate $ZnCO_3(s)$ or $ZnO(s)$ can deposit on soil surface and further become physically entrapped by soil particles. Mineral and humic constituents of soils also provide functional surface groups that can interact more directly with metal ions by 1) surface complexation, and 2) adsorption followed by surface precipitation (25).

Complexation and Release of Zinc by ADA. In order to be effective in extracting zinc from soil and forming a soluble complex, ADA must overcome competing processes including 1) the metal's precipitation as hydrous oxide and/or carbonate, and 2) surface complexation and precipitation on soil particle surface. Equilibrium reactions and associated constants needed to assess the complexation ability of ADA toward zinc are listed in Table II. For a system of 1 mM Zn_T and 10-fold excess

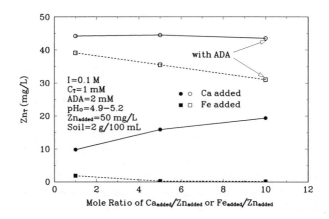

Figure 5   Adsorption (solid symbols) of zinc to soil and extraction (open symbols) with ADA in the presence of competing Fe(III) and Ca ions.

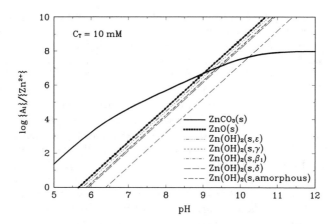

Figure 6   Log activity ratio diagram for various zinc oxide, hydrous oxide, and carbonate solids, indicating their relative stability over a wide pH range.

of ligand, Figure 7 shows the degrees of complexation, as log Zn_T/Zn_T' (26), for ADA toward Zn. The term Zn_T is defined as:

$$Zn_T = Zn_T' + ZnADA^0 + Zn(ADA)_2^{2-} + Zn(H_{-1}ADA)ADA^{3-} + Zn(H_{-1}ADA)_2^{4-} \quad (4)$$

where Zn_T' is defined as:

$$Zn_T' = [Zn^{2+}] + [ZnOH^+] + [Zn(OH)_2^0(aq)] + [Zn(OH)_3^-] + [Zn(OH)_4^{2-}]$$

$$+ [Zn_2OH^{3+}] + [Zn_4(OH)_4^{4+}] + [ZnCO_3^0(aq)] + [ZnHCO_3^+] \quad (5)$$

Also shown in Figure 7 are degrees of complexation for other ligands toward Zn including EDTA, NTA, citric acid, trans-1,2-diaminocyclohexanetetraacetic acid (CDTA), and ethylene-N,N'-bis(o-hydroxyphenylglycine) (EHPG). As the degree of complexation curve for ADA shows, a 10-fold excess of ADA is able to maintain 1 mM Zn completely soluble in aqueous solution for any pH value below 10.5. EDTA, NTA, CDTA, and EHPG are more effective chelating agents in solubilizing zinc. However, only ADA is predicted to release zinc once the solution pH is to be raised to 10.5, as indicated by an intersection between the curve with the chelator and the Zn precipitate curve without the chelator. The intersection (pH 10.3) is better defined when 10 mM of Ca^{II} is added. While all other chelators shown are equally or more powerful than ADA in complexing with zinc, they are not predicted to release the metal readily, thus the recovery of metal and reuse of chelator may be difficult.

It should be noted that the initial pH of most adsorption experiments was between 5 and 6 to ensure that the zinc solution (Zn_{added} = 50 ppm) was homogeneous prior to soil addition. However, the solution pH typically increased slightly as adsorption or extraction progressed due to the buffering action (acid-neutralizing capacity) of the soil. For example, the solution pH increased from pH_0 = 5.1 to pH's 6.3 and 6.6 after 24 h of equilibration with 2.2 g/L soil and 23 g/L soil, respectively. As expected, the extent of shift in pH depended on the amount of soil used and the initial acidity of the solution. In this study, limited experiments were performed to address the effect of conditions possibly encountered in field application, such as the presence of competing ions, and the stability of the chelator under extended contact with soil. Figures 5 and 3 show that ADA prefers Zn over Fe and Ca, and that ADA in contact with the soil remains stable over an one-month test period, respectively. However, more experiments with varying conditions should be performed to assess these factors fully. Further, it should be noted that although this study successfully tested the stated objectives using a laboratory spiked soil, the aging of Zn metal in the spiked soil was insignificant compared to that in an authentic contaminated soil. Therefore, the extraction and recovery of zinc from authentic contaminated soils should be tested. Work is in progress in our laboratories using authentic contaminated and our finding will be presented in forthcoming reports.

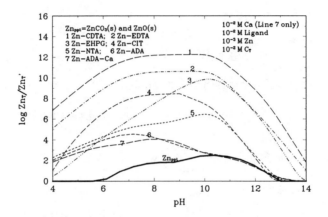

Figure 7   Degrees of complexation of ADA and other chelators toward zinc as a function of pH.

ADA   ($C_6H_{10}O_5N_2$) N-2-acetamidoiminodiacetic acid;
CDTA   ($C_{14}H_{22}O_8N_2$) Trans-1,2-diaminocyclohexanetetraacetic acid;
CIT   ($C_6H_8O_7$) Citric acid;
DTPA   ($C_{14}H_{23}O_{10}N_3$) Diethylenediaminepentaacetic acid;
EDTA   ($C_{10}H_{16}O_8N_2$) Ethylenediaminetetracetic acid;
EGTA   ($C_{14}H_{24}O_{10}N_2$) Ethyleneglycolbis(ethylamine)tetraacetic acid;
EHPG   ($C_{18}H_{20}O_6N_2$) Ethylene-N,N'-bis(o-hydroxyphenylglycine);
NTA   ($C_6H_9O_6N$) Nitrilotriacetic acid.

Conclusion

Using a laboratory Zn-spiked soil, this work demonstrated that 1) the chosen chelator ADA was able to selectively extract Zn over Fe and Ca from soil; 2) the extracted zinc was recovered as a ZnO precipitate by increasing the solution pH; and 3) the recovered ADA was amenable to reuse during subsequent extraction runs. The work suggests ADA a potential remediating agent for soils contaminated with Zn, and the results warrant further investigation on using ADA for extraction of Zn and other heavy metals from authentic contaminated soils and design for pilot studies.

Symbols

C_T total carbonate concentration (M)
$$C_T = [H_2CO_3] + [HCO_3^-] + [CO_3^{2-}]$$

Zn_T total dissolved zinc concentration (M, mg/L)

Zn_T' total dissolved zinc concentration excluding the ligand complexed (M, mg/L).

Zn_{added} total amount of zinc added to the system (mg/L)

I ionic strength (M)

L_T total ligand (chelator) concentration, including all conjugate acids and bases and the complexed ligands (M)

V volume (mL)

Literature Cited

1. Sims, R. C. *J. Air Waste Manage. Assoc.* **1990**, *40*, 704-732.
2. *The Superfund Innovative Technology Evaluation Program: Technology Profiles*; U.S. Environmental Protection Agency. Office of Research and Development. Risk Reduction Engineering Laboratory. Washington, DC, 1990; EPA/540/5-90/006.
3. *Synopses of Federal Demonstrations of Innovative Site Remediation Technologies;* Member Agencies of the Federal Remediation Technologies Roundtable. 1991; EPA/540/8-91/009.
4. Peters, R. W.; Shem, L. *Environ. Progress* **1992**, *11*, 234-240.
5. Slavek, J.; Pickering, W. F. *Water, Air, and Soil Pollution* **1986**, *28*, 151-162.
6. Elliott, H. A.; Linn, J. H.; Shields, G. A. *Hazardous Waste & Hazardous Materials* **1989**, *6*, 223-229.

7.   Ellis, W. D.; Fogg, T. R.; Tafuri, A. N. *Treatment of Soils Contaminated with Heavy Metals* In Proceedings of the 12th Annual Research Symposium: Land Disposal, Remedial Action, Incineration and Treatment of Hazardous Waste. **1986**, 201-207.
8.   Borggaard, O. K. *J. Soil Sci.* **1979**, *30*, 727-734.
9.   Elliott, H. A.; Huang, C. P. *J. Colloid and Interface Sci.* **1979**, *70*, 29-45.
10.   Sommers, L. E.; Lindsay, W. L. *Soil Sci. Soc. Am. J.* **1979**, *43*, 39-47.
11.   Francis, A. J.; Dodge, C. J.; Gillow, J. B. *Nature* **1992**, *356*, 140-142.
12.   Huang, C. P.; Rhoads, E. A.; Hao, O. J. *Wat. Res.* **1988**, *22*, 1001-1009.
13.   Bowers, A. R.; Huang, C. P. *J. Colloid and Interface Sci.* **1986**, *110*, 575-590.
14.   Hohl, H.; Stumm, W. *J. Colloid and Interface Sci.* **1976**, *55*, 281-288.
15.   Vuceta, J.; Morgan, J. J. *Environ. Sci. Technol.* **1978**, *12*, 1302-1309.
16.   Davis, J. A.; Leckie, J. O. *Environ. Sci. Technol.* **1978**, *12*, 1309-1315.
17.   McLean, E. O. In *Methods of Soil Analysis*, 2nd ed.; Page, A. L.; Miller, R. H.; Keeney, D. R., Eds.; American Society of Agronomy and Soil Science Society of American: Madison, WI, 1982, Part 2; pp 199-224.
18.   Raij, B. V.; Peech, M. *Soil Sci. Soc. Amer. Proc.* **1972**, *36*, 587-593.
19.   Das, B. M. In *Soil Mechanics Laboratory Manual*, 2nd ed. Engineering Press: San Jose, CA, 1986; 9-32.
20.   Rhoades, J. D. In *Methods of Soil Analysis*, 2nd ed.; Page, A. L.; Miller, R. H.; Keeney, D. R., Eds.; American Society of Agronomy, and Soil Science Society of American: Madison, WI, 1982, Part 2; pp 149-157.
21.   Chang, A. C.; Page, A. L.; Warneke, J. E.; Grgurevic, E. *J. Environ. Qual.* **1984**, *13*, 33-38.
22.   *ASTM Annual Standards* (Annual) American Society of Testing and Materials, Philadelphia.
23.   *Standard Methods for the Examination of Water and Wastewater*; 17th ed., Washington, American Public Health Association: Washington, DC, 1989.
24.   Martell A. E.; Smith R. M. *Critical Stability Constants*; Vol. 1, 4, 5, and 6; Plenum; New York, 1974, 1976, 1982, 1989.
25.   Dzombak, D. A.; Morel F. M. M. *Surface Complexation Modeling of Hydrous Ferric Oxide*; John Wiley & Sons; New York, 1990.
26.   Stumm, W.; Morgan, J. J. *Aquatic Chemistry*; 2nd ed.; John Wiley & Sons: New York, 1981.

RECEIVED March 14, 1995

Chapter 18

Enhanced Removal of Organic Contaminants by Solvent Flushing

D. C. M. Augustijn[1] and P. S. C. Rao

Soil and Water Science Department, University of Florida, Gainesville, FL 32611

The use of cosolvents to enhance the remediation of contaminated soils is based on four observations: (1) enhanced mobilization of a residual NAPL phase; (2) increased solubility; (3) reduced sorption or retardation; and (4) increased mass-transfer rates. The theoretical basis for each of those mechanisms is briefly reviewed. Solvent washing technologies for treatment of excavated soils are already commercially available. The potential use of organic cosolvents for in situ remediation (solvent flushing) is illustrated by a review of several studies that showed an enhanced removal of contaminants upon addition of a cosolvent. Possible problems that may arise when cosolvents are used for in situ remediation are indicated, which should be helpful for planning further research and to develop solvent flushing into a viable remediation technique.

With a growing experience in site remediation, it becomes more evident that many conventional technologies are not able to restore contaminated sites to required clean-up levels in a reasonable time frame. This fact has stimulated the search for alternative technologies that are able to clean up contaminated sites in a timely and cost-effective way. Several new technologies are being developed, among which the use of various adjuvants is receiving considerable attention. There are essentially two types of adjuvants: (1) those that enhance the release and mobility of contaminants (e.g., cosolvents, surfactants), and (2) those that enhance the transformation of contaminants (e.g., nutrients, ozone, chemical reaction agents). This paper will focus on the first type of adjuvants, in particular on the use of cosolvents for enhanced remediation of source areas at waste disposal sites.

[1]Current address: Department of Civil Engineering and Management, University of Twente, P.O. Box 217, 7500 AE Enschede, Netherlands

Cosolvency Theory

The use of cosolvents (e.g., alcohols) for soil remediation is based on four principles. First, cosolvents decrease the interfacial tension between the solution phase and a nonaqueous phase liquid (NAPL), inducing mobilization of a trapped NAPL phase. Second, cosolvents increase the solubility of non-polar organic chemicals, enhancing the release of organic constituents of an immobile NAPL phase. Third, cosolvents reduce sorption, facilitating faster transport of dissolved contaminants. Fourth, mass-transfer rates generally increase in the presence of cosolvents. These characteristics all favor an enhanced removal of organic chemicals from contaminated soils, and therefore indicate a potential application of cosolvents for remediation of contaminated sites. In the following sections, the cosolvency theory will be briefly reviewed.

Mobilization. Many industrial waste disposal/spill sites are contaminated with immiscible nonaqueous phase liquids (NAPLs). If the NAPL is present as a free moving phase, a part of it can be recovered hydraulically. What is left behind is a residual saturation of discontinuous ganglia trapped by capillary forces in the pore spaces. The ratio of viscous and capillary forces acting on the residual NAPL is expressed by the capillary number:

$$N_c = \frac{\mu\,v}{\phi} \tag{1}$$

where μ is the dynamic viscosity of the solution phase, v is the groundwater velocity, and ϕ is the interfacial tension between the solution and NAPL phase. In general, residual NAPL is mobilized as the capillary number is greater than 2×10^{-5} and virtually all NAPL is displaced when the capillary number exceeds 5×10^{-3} (*1*). Equation 1 indicates that mobilization of the NAPL globules can be established by increasing the groundwater velocity, decreasing the NAPL-water interfacial tension, or a combination of both. Given the practical limitations on groundwater pumping rates, reducing the interfacial tension by adding cosolvents or surfactants to groundwater may be a more viable option for remediating sites contaminated with a residual NAPL phase.

Cosolvents may enhance the mobilization of a residual NAPL phase also as a result of the partitioning of the cosolvent into the NAPL, which depends on the hydrophobicity of the NAPL as well as the cosolvent. When a cosolvent partitions preferentially into the NAPL phase, the residual NAPL globules may swell considerably. The swollen NAPL globules may become a relative continuous phase, making them much easier to displace. In addition, the swelling will reduce the density of the NAPL phase, which increases the buoyancy force acting on submerged NAPL globules. This makes it easier to displace the residual NAPL in an upward direction and reduces the potential for further downward migration (*2*).

Dissolution. NAPLs may consist of one or more components. Chlorinated solvents like tri- and perchloroethylene (TCE, PCE) are prominent examples of single-component NAPLs. Gasoline, motor oil, diesel, creosote, and coal tar are typical examples of multi-component NAPLs. Due to the limited solubility of most organic constituents in water, removal of the NAPL by dissolution in water, as in pump-and-treat techniques, often requires years or even decades (3). By adding an organic cosolvent to water, the polarity of the solvent mixture decreases, resulting in an increased solubility of non-polar organic chemicals. The solubility of a non-polar organic solute in a binary solvent mixture (S_m) increases in a nearly log-linear manner with increasing volume fraction of cosolvent (f_c) (4–6):

$$\log S_m = \log S_w + \beta \sigma f_c \qquad (2)$$

where

$$\sigma = \log \frac{S_c}{S_w} \qquad (3)$$

β is an empirical coefficient that accounts for water-cosolvent interactions, S_c is the solubility in the neat organic solvent, S_w is the solubility in water, and σ is the cosolvency power of the organic solvent for the solute of interest. When it is assumed that the cosolvent effects are additive, equation 2 can be generalized as follows for a mixture of water and several cosolvents:

$$\log S_m = \log S_w + \sum_{i=1}^{N} \beta_i \sigma_i f_{c,i} \qquad (4)$$

where the subscript i designates the values for the ith cosolvent, and N the number of cosolvents in the mixture.

The cosolvency power (σ) is an important parameter in the log-linear model since it indicates the capacity of a cosolvent to increase the solubility of an organic solute. The cosolvency power will increase with increasing hydrophobicity of the solute as well as the solvent. Since the degree of hydrophobicity for solutes is much more variable than for solvents, σ values can be estimated from polarity or hydrophobicity indices for the solute, such as the octanol-water partition coefficient (6).

Non-ideal behavior due to water-cosolvent interactions generally lead to positive deviations from the log-linear model. For small deviations, the empirical coefficient β (≥ 1) can be used to predict the solubility over a range of cosolvent fractions. However, when water-cosolvent interactions are significant, more sophisticated methods should be used to predict activity coefficients in the solvent mixture (7, 8). Also, when the NAPL is completely dissolved in the mixed solvent, particular at high cosolvent fractions, the log-linear approximation needs to be replaced by a more general approach based on phase diagrams.

For groundwater in contact with a single-component NAPL, the equilibrium concentration is equal to the aqueous solubility. The increase in solubility, and hence equilibrium concentration, upon addition of a cosolvent is predicted by equation 4.

The equilibrium concentration (*C*) of component *i* in a solution which is in contact with a multi-component NAPL can be predicted by:

$$C_i = \gamma_i\, X_i\, S_i \qquad (5)$$

where γ_i is the activity coefficient of component *i* in the NAPL phase, X_i is the mole fraction in the NAPL phase, and S_i is the aqueous solubility of the pure component in its liquid state (i.e., for solids, the solubility of a hypothetical, super-cooled liquid is considered). In many cases the activity coefficient of an organic constituent in an organic NAPL can be assumed unity, simplifying equation 5 to Raoult's law.

Since the solubility increases in a log-linear manner with increasing cosolvent content, it follows that C_i obeys the same log-linear model. Thus, cosolvent addition to a multi-component waste mixture can be expected to result in an exponential increase in solution-phase concentrations of organic constituents of the waste. This prediction is consistent with the data reported by Lane and Loehr (*9*) for the dissolution of several polycyclic aromatic hydrocarbons from tar-contaminated soils into binary mixtures of alcohols and water. It should be noted, however, that the activity coefficient and mole fraction of NAPL components can change if a significant amount of cosolvent partitions into the NAPL phase. It is also possible to describe the dissolution of a multi-component NAPL in a solvent mixture by a single pseudo-component (*10*).

Sorption. Sorption of non-polar organic compounds can generally be described by a linear sorption isotherm, characterized by an equilibrium partition coefficient (*K*). Since sorption is inversely related to the solubility, an increase in solubility, resulting from addition of cosolvents, leads to a proportional decrease in sorption. Thus, the equilibrium sorption coefficient (K_m) measured in mixed solvents decreases in a log-linear manner with increasing cosolvent content (*11–13*):

$$\log K_m = \log K_w - \sum_{i=1}^{N} \alpha_i \beta_i \sigma_i f_{c,i} \qquad (6)$$

where α is an empirical coefficient that accounts for solvent-sorbent interactions, the subscripts *m* and *w* denote mixed solvent and water, respectively, and all other terms are as defined previously. The reduction in sorption upon addition of a cosolvent will enhances the mobility of organic solutes.

Non-ideal behavior due to cosolvent-sorbent interactions may result in either positive deviations ($\alpha > 1$) or negative deviations ($\alpha < 1$). The product $\alpha\beta$ accounts for deviations arising from both cosolvent-water and cosolvent-sorbent interactions. Values for α and β vary with cosolvent, soil, and chemical. Since limited information is available for estimating α or β values for specific cases, as a first approximation, the product $\alpha\beta$ can be assumed equal to 1.

Nonequilibrium Conditions. In many field-scale applications, equilibrium approaches prove to be inadequate in estimating the dissolution or sorption processes during transport since nonequilibrium conditions usually prevail. Failure of pump-and-treat, soil venting or bioremediation techniques has been attributed to nonequilibrium conditions at all spatial scales. There are essentially two types of nonequilibrium conditions. The first type results from the heterogeneous character of porous media that causes the hydraulic conductivity to be spatially variable, and results in a heterogeneous distribution of groundwater velocities. The concentration gradients created by the nonuniform velocity distribution can result in rate-limited diffusion of the solute between different flow domains, causing nonequilibrium conditions. This type of nonequilibrium is often referred to as physical nonequilibrium. The second type of nonequilibrium occurs when the contact time between phases is not sufficient to achieve an equilibrium distribution of the chemicals in all phases. Typical examples of this type of nonequilibrium conditions are dissolution and sorption nonequilibrium. In the following section, we will briefly examine the effects of cosolvents on rate-limited processes.

Physical Nonequilibrium. It is known that organic solvents, such as alcohols, have the potential to increase the intrinsic permeability of soils and aquifers, especially in media with high clay contents (*14, 15*). This characteristic is due to shrinking and cracking of the clay as a result of the low dielectric constant of cosolvents, and may have a positive effect on the recovery of contaminants from zones of low permeability in at least two ways. First, cracking will increase the advective flow through the clay lenses which are otherwise accessed only via molecular diffusion. Second, the average diffusion path length is decreased as a result of fissuring or cracking of the clay lenses. On the other hand, the cosolvent still has to diffuse into and the contaminants have to diffuse out of the denser clay regions developed in between the cracks. Data are lacking to estimate the extent to which the enhanced permeation is offset by decreased diffusion rates.

Dissolution Nonequilibrium. Dissolution processes are generally considered to be limited by diffusion of the component through a boundary layer from the interface into the bulk solution (*16*). The rate of dissolution from a NAPL ($\partial m/\partial t$) is often described by a first-order rate law:

$$\frac{\partial m}{\partial t} = -\theta \, k (C_{eq} - C) \tag{7}$$

where m is the contaminant mass in the NAPL phase, t is the time, θ is the volumetric water (or mixed solvent) content, k is the first-order dissolution rate coefficient, and C is the solution-phase concentration. C_{eq} is the equilibrium concentration which equals the solubility for single-component NAPLs or the equilibrium concentration based on Raoult's law (equation 5) for multi-component NAPLs. The mass-transfer coefficient for dissolution (k) is often expressed as the dimensionless Sherwood number and appears to have positive correlation with pore-water velocity, NAPL saturation, and fluid properties such as viscosity and density (*17*). It is probable that interactions of the NAPL with natural soil organic matter will also modify the dissolution rates.

When a cosolvent is added to a soil contaminated with a NAPL, the solubility, and hence C_{eq}, will increase according to equation 4, enhancing the dissolution rate. In addition, the changes in fluid properties may also affect the dissolution rate. In viscous, multi-component NAPLs, like coal tar or crude oil, diffusive transport within the organic phase is most likely the mechanism controlling the dissolution kinetics (*18*). When the dissolution process proceeds, the low molecular weight components will dissolve while the large molecular weight components remain, making the NAPL phase more viscous and the diffusion even slower. The partitioning of cosolvents into the NAPL phase may prevent this to some extent. So far, very little information is available on the effect of cosolvents on dissolution kinetics. More research is needed to accurately predict the dissolution rate of NAPLs in solvent mixtures.

Sorption Nonequilibrium.   Sorption nonequilibrium has generally been described by a bicontinuum model where sorption is considered to take place in two steps. The first step is considered to occur instantaneously in domains that are readily accessible for the sorbate (S_1), while the second step is rate-limited, often attributed to a diffusion controlled process (S_2). Some researches have proposed retarded intra-particle diffusion as a mechanism to describe rate-limited sorption (*19, 20*), while others have described sorption nonequilibrium by intra-organic matter diffusion (*21, 22*). The bicontinuum model can be conceptualized as follows:

$$C \xleftrightarrow{FK} S_1 \xleftrightarrow{\lambda} S_2 \tag{8}$$

where F is the fraction of sorption occurring in the instantaneous regions, and λ is a first-order mass-transfer rate coefficient that describes the rate-limited sorption. Several researchers found an increase in sorption rate coefficients with increasing cosolvent fractions (*23–25*). Brusseau et al. (*25*) attributed this to the swelling of organic matter in the presence of an organic cosolvent, thus enhancing sorbent permeation by the sorbate. In case of retarded intra-particle diffusion, this phenomenon can be explained by reduced retardation of the solute in the micropores of porous particles.

Applications

Based on the cosolvency theory described above, it is expected that cosolvents can be useful for remediating source areas of waste disposal/spill sites contaminated with organic chemicals. Interest in the use of solvents to facilitate the removal of hydrocarbons is not new. In the early 1960s solvent flushing was investigated by petroleum engineers as a technique to increase the recovery of oil from deep reservoirs (*26–29*). Although various alcohols were found to be capable of enhancing oil displacement, the solvent flushing technique never found wide application in the oil industry because of the relative high cost of alcohols compared to the increased recovery. Since recently, there is a renewed interest in the use of solvents for environmental application. The limitations of most conventional remediation technologies has generated a search for new and innovative technologies that are able

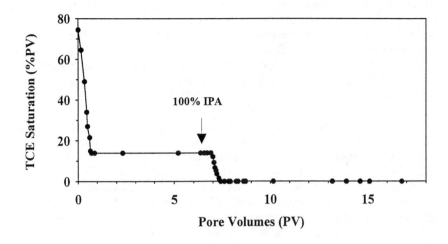

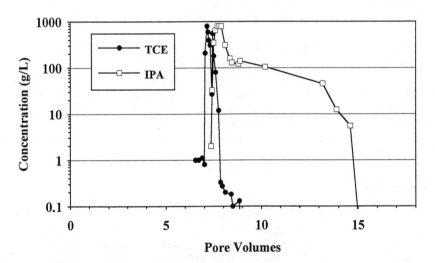

Figure 1. Displacement of TCE by one pore volume of 100% isopropyl alcohol. (Reproduced with permission from ref. 30. Copyright 1992 Lewis Publishers.)

to clean up contaminated sites more effectively.   Solvent flushing is one of the emerging technologies that seems to be an attractive alternative.

Currently, solvents are used on a commercial basis for treatment of excavated soils (solvent washing).   The two most common techniques are the Basic Extraction Sludge Treatment (BEST) and the CF Systems process.   In the BEST process, triethylamine (TEA) is used as the solvent to extract organic chemicals from contaminated media.   TEA is miscible with water below 15°C, and separates from water above this temperature.   For treatment, the contaminated soil or sludge is mixed with TEA in a reactor.   The contaminants are extracted into the liquid phase, which is then separated from the soil and heated to separate the TEA from the water phase, so that the TEA can be recycled.   The CF Systems process uses propane or carbon dioxide at super-critical conditions (high temperature and high pressure).   In this state, the solvent has the extraction capacity of a liquid, and the diffusive properties of a gas.   After mixing with the soil, the super-critical fluid is separated again, and the pressure reduced.   At this point the solvent becomes a gas, while the extracted organics remain liquid.   The solvent is then compressed and recycled.

Remediation of contaminated soils using ex situ technologies requires excavation and sophisticated treatment equipment, making it very expensive.   In addition, excavation increases the risk of exposure, and requires a significant destruction of the landscape.   In many cases, in situ treatment of contaminated sites may therefore be preferred.   The use of solvents for in situ remediation (solvent flushing) has been demonstrated in several laboratory column experiments of which some will be reviewed below.

Laboratory Experiments.   Boyd and Farley (*30*) used isopropyl alcohol (IPA) to study the enhanced mobilization of residual trichloroethylene (TCE) globules in column packed with glass beads.   Solutions of 100 and 50% isopropyl alcohol were able to displace the immiscible TCE within one pore volume (Figure 1a).   A solution of 10% isopropyl alcohol did not reduce the interfacial surface tension sufficiently to mobilize the TCE globules, but showed an increase in effluent concentration of TCE as a result of increased solubility.   Boyd and Farley (*30*) also observed the instability (fingering) of the mixed-solvent front due to differences in density and viscosity. Figure 1b, for example, shows that it takes several pore volumes of water to displace the one pore volume of isopropyl alcohol that was injected to mobilize the TCE in the downflow direction of a vertical column.   To avoid instabilities at the front a gradient injection of the cosolvent may be considered.

Rixey et al. (*31*) compared water, solvent and surfactant flushing for the removal of a synthetic hydrocarbon mixture from a sandy soil.   They concluded that solvent and surfactant flushing were far more effective in removing the hydrocarbon than water flushing.   Approximately 300 g of pure methanol was required to remove the 27 g of hydrocarbon mixture initially applied to the column by mobilization and solubilization, about a thousand times more effective than removal by water.   Rixey et al. (*31*) also observed that the release of a small fraction of the mass (< 0.01%) was rate limited.

Augustijn et al. (*32*) conducted column experiments with a Eustis fine sand that was initially equilibrated with an aqueous naphthalene solution, and then eluted with

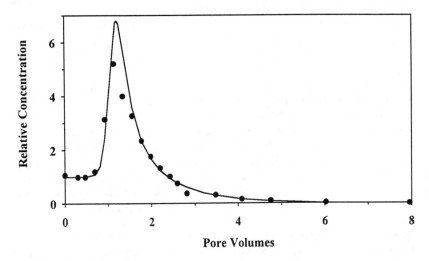

Figure 2. Experimental and predicted elution profiles for a naphthalene contaminated soil flushed with 30% methanol. (Reproduced with permission from ref. 32. Copyright 1994 ASCE.)

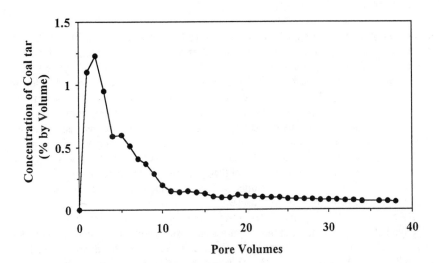

Figure 3. Elution profile for a glass bead column containing a 20% initial saturation of coal tar flushed with 80% n-butylamine. (Reproduced with permission from ref. 33.)

methanol-water mixtures to remove the naphthalene from the sorbed phase. Model simulations based on independently estimated parameters showed good agreement with the measured elution profiles (Figure 2). In general, removal efficiency increased with increasing cosolvent content, and decreasing hydrophobicity of the solute.

Roy (*33*) and Augustijn (*18*) studied the release of organic constituents from coal tar contaminated soils in water/cosolvent systems. Coal tar is a complex waste mixture that consists primarily of polycyclic aromatic hydrocarbons. Roy (*33*) conducted batch and column experiments to investigate the removal of coal tar from glass beads and a sandy soil with n-butylamine. Figure 3 gives an example of a typical elution profile. Note that Roy considered coal tar as a single pseudo-component. Augustijn (*18*) monitored the concentration of individual coal tar components in the effluent of solvent flushing experiments. The elution profiles were predicted based on Raoult's law and the cosolvency theory. Figure 4 shows the results of naphthalene and fluorene for flushing with water and a 50% methanol solution. For fluorene, an increase in concentration can be observed, which is a result of an initial increase in the mole fraction of fluorene as low molecular weight components dissolve from the tar. Both studies showed that cosolvents enhanced the removal of organic constituents, but in general many pore volumes (> 10) were required to remove the soluble fraction from the coal tar. It is uncertain, however, what fraction of the coal tar needs to be removed to reduce the environmental risk sufficiently.

Field Application. To date, there has been at least one case in which the solvent flushing technique has been tested in the field. Broholm and Cherry (*34*) carried out a field experiment in which 5 liters of a ternary mixture of NAPLs (10% chloroform, 40% TCE, and 50% PCE) was introduces in a 57 m^3 hydraulically isolated test cell in a sandy aquifer. Several months after the NAPL mixture was released, a 5.5 day pulse of a 30% methanol solution was injected. The concentration of TCE and PCE near the extraction wells increased by a factor of 2 to 5 at the same time as methanol appeared. This increase was larger than that expected based on the methanol concentration likely to be present in the NAPL-contaminated zone, suggesting that some NAPL mobilization may have occurred. Later excavation of the test cell showed that 30% of the NAPL mixture was removed during the entire experiment.

Although cosolvents may have a potential application for in situ remediation, several problems still need careful consideration before implementation in the field. Such problems include: (1) delivery of the additive to the source; (2) reduced efficiency of the flushing treatment due to soil heterogeneities and non-uniform distribution of the contaminants; (3) transport of dissolved contaminants off site; (4) recovery and waste management; and (5) costs. The choice of a suitable solvent depends on the extraction capacity of the solvent (cosolvency power), biodegradability, and cost. The solvent must have favorable density, viscosity, and interfacial tension properties. In some cases, combination of cosolvents may be considered to optimize the properties of the solvent mixture. There should also be a sufficient difference between the boiling points of the solute, water, and solvent to enable post-treatment separation. Solvent treatment will most likely be suitable for hydrocarbons with low vapour pressure, since less costly remediation techniques are available for volatile compounds.

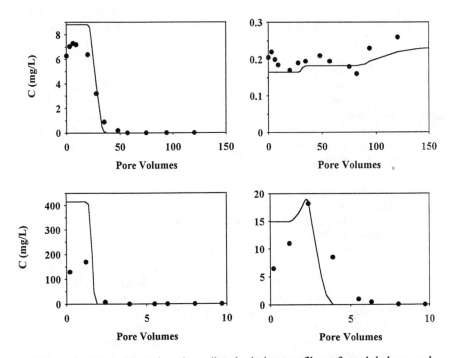

Figure 4. Experimental and predicted elution profiles of naphthalene and fluorene for a coal tar contaminated soil flushed with water and 50% methanol. (Adapted from ref. 18.)

Conclusions

In this paper, we presented a brief overview of the cosolvency theory. Based on the properties of organic solvents, it was suggested that cosolvents can be used as chemical adjuvants to enhance the remediation of contaminated soils. Solvent washing technologies for treatment of excavated soils are already commercially available. The potential use of organic solvents for in situ remediation (solvent flushing) has been illustrated by a review of several studies that showed an enhanced removal of contaminants upon addition of a cosolvent. Although these studies indicate a potential application of cosolvents for in situ remediation, more experimental data are needed to confirm these studies, and to work out the problems that may be associated with the implementation of the solvent flushing technique at the field scale. This review should be helpful in planning further research and to develop the solvent flushing technique into an environmentally safe and economically attractive remediation alternative.

Acknowledgements

Research support from the U.S. Environmental Protection Agency (CR-814512) and the Electric Power Research Institute (RP-2879-7) is greatfully acknowledged. Although the research described in this article has been funded by the U.S. EPA through a cooperative agreement, it has not been subjected to the agency's required peer and policy review, and therefore does not necessarily reflect the views of the agency and no official endorsement should be inferred.

Literature Cited

(1)   Hunt, J. R.; Ditar, N.; Udell, K. S. *Water Resour. Res.* **1988**, *24*, 1247.
(2)   Brandes D.; Farley, K. J. *Water Environ. Res.* **1994** (in press).
(3)   Mackay, D. M.; Cherry, J. A. *Environ. Sci. Technol.* **1989**, *23*, 630.
(4)   Yalkowsky, S. H.; Roseman, T. In *Techniques of Solubilization of Drugs*; Yalkowsky, S. H., Ed.; Marcel Dekker: New York, NY, 1981; pp 91-134.
(5)   Fu, J. K.; Luthy, R. G. *J. Environ. Eng.* **1986**, *112*, 328-345.
(6)   Morris, K. R.; Abramowitz, R.; Pinal, R.; Davis, P.; Yalkowsky, S. H. *Chemosphere* **1988**, *17*, 285-298.
(7)   Groves, F. R. *Environ. Sci. Technol.* **1988**, *22*, 282-286.
(8)   Pinal, R.; Lee, L. S.; Rao, P. S. C. *Chemosphere* **1991**, *22*, 939-951.
(9)   Lane, W. F.; Loehr, R. C. *Environ. Sci. Technol.* **1992**, *26*, 983-990.
(10) Peters, C. A.; Luthy, R. G. *Environ. Sci. Technol.* **1993**, *27*, 2831-2843.
(11) Rao, P. S. C.; Hornsby, A. G.; Kilcrease, D. P.; Nkedi-Kizza, P. *J. Environ. Qual.* **1985**, *14*, 376-383.

(12) Nkedi-Kizza, P.; Rao P. S. C.; Hornsby, A. G. *Environ. Sci. Technol.* **1985**, *19*, 975-979.
(13) Fu, J. K.; Luthy, R. G. *J. Environ. Eng.* **1986**, *112*, 346-366.
(14) Schramm, M.; Warrick, A. W.; Fuller, W.H. *Hazard. Wastes Hazard. Mat.* **1986**, *3*, 21-27.
(15) Brown, K. W.; Thomas, J. C. *Soil Sci. Soc. Am. J.* **1987**, *51*, 1451-1459.
(16) Pfannkuch, H. O. *U.S. Geol. Surv. Water Resour. Invest. Rep.* **1984**, *84-4188*, 23-48.
(17) Powers, S. E.; Loureiro, C. O.; Abriola, L. M.; Weber, W. J. *Water Resour. Res.* **1991**, *27*, 463-477.
(18) Augustijn, D. C. M. Ph. D. Dissertation, Soil and Water Science Department, University of Florida, Gainesville, FL, 1993.
(19) Wu, S.; Gschwend, P. M. *Environ. Sci. Technol.* **1986**, *20*, 717-725.
(20) Ball, W. P.; Roberts, P.V. *Environ. Sci. Technol.* **1991**, *25*, 1237-1249.
(21) Karickhoff, S. W.; Morris, K. R. *Environ. Toxic. Chem.* **1985**, *4*, 469-479.
(22) Brusseau, M. L.; Jessup, R. E.; Rao, P. S. C. *Environ. Sci. Technol.* **1991**, *25*, 134-142.
(23) Nkedi-Kizza, P.; Brusseau, M. L.; Rao, P. S. C.; Hornsby, A. G. *Environ. Sci. Technol.* **1989**, *23*, 814-820.
(24) Wood, A. L.; Bouchard, D. C.; Brusseau, M. L.; Rao, P. S. C. *Chemosphere* **1990**, *21*, 575-587.
(25) Brusseau, M. L.; Wood, A. L.; Rao, P. S. C. *Environ. Sci. Technol.* **1991**, *25*, 903-910.
(26) Gatlin, C.; Slobod, R. L. *Trans AIME* **1960**, *219*, 46-53.
(27) Taber, J. J.; Kamath, I. S. K.; Reed, R. L. *Soc. Petrol. Eng. J.* **1961**, *Sept.*, 129-142.
(28) Holm, L. W.; Csaszar, A. K. *Soc. Petrol. Eng. J.* **1962**, *June*, 129-142.
(29) Taber, J. J.; Meyer, W. K. *Soc. Petrol. Eng. J.* **1964**, *March*, 37-48.
(30) Boyd, G. R.; Farley, K. J. In *Hydrocarbon Contaminated Soils and Groundwater*; Calabrese, E. J.; Kostecki, P. T., Eds.; Lewis Pulishers: Chelsea, MI, 1992, Vol. 2, pp 437-460.
(31) Rixey, W. G.; Johnson, P. C.; Deeley, G. M.; Byers, D. L.; Dortch, I. J. In *Hydrocarbon Contaminated Soils and Groundwater*; Calabrese, E. J.; Kostecki, P. T., Eds.; Lewis Pulishers: Chelsea, MI, 1991, Vol. 1, pp 387-409.
(32) Augustijn, D. C. M.; Jessup, R. E.; Rao, P. S. C.; Wood, A. L. *J. Environ. Eng.* **1994**, *120*, 42-57.
(33) Roy, S. B. M. Sc. Thesis, Department of Civil Engineering, Carnegie Mellon University, Pittsburgh, PA, 1992.
(34) Broholm, K.; Cherry, J. A. In *Transport and Reactive Processes in Aquifers*; Dracos, T.; Stauffer, F., Eds.; Balkema: Rotterdam, the Netherlands, 1994, pp 563-568.

RECEIVED March 14, 1995

Chapter 19

Hot Water Enhanced Remediation of Hydrocarbon Spills

Eva L. Davis

R. S. Kerr Environmental Research Laboratory, U.S. Environmental Protection Agency, P.O. Box 1198, Ada, OK 74820

When viscous, nonvolatile oils are contaminating the subsurface, hot water may be used to lower the viscosity of the oil and increase its mobility. The purpose of this research is to demonstrate the use of moderately hot water to enhance the remediation of waste oils from soils or aquifers. Displacement experiments have been performed at constant temperatures using two different oils and two sands. These experiments have shown that as the temperature increases, oil recovery also increases, reducing the oil content by as much as 33 percent. Transient temperature experiments performed with the column held at 10°C and using 50°C water to displace the oil showed that the benefits of hot water displacement can be achieved under conditions which more closely resemble field conditions.

Many different types of immiscible contaminants have been identified at polluted sites, and the wide range of properties of these contaminants means that a variety of remediation technologies will be required for the efficient and cost effective cleanup of these sites. Gasoline, jet fuel, crude oil, chlorinated organic solvents, wood preserving chemicals, and transmission fluids are some of the immiscible fluids that have been found as free organic phases contaminating the subsurface. Some of these oils, such as crude oils, wood preserving chemicals, or transmission fluids, can present special problems in remediation efforts due to their high viscosity, low solubility in water, and low volatility. These oils are not amenable to pump and treat remediation or processes which remove contaminants in the gas phase such as vacuum extraction or air sparging. For contaminants such as these, heat in the form of hot water may be used to lower the viscosity and thereby increase the movement of the oil to a recovery well or trench.

The use of hot water to lower the viscosity and increase oil mobility has been used by the oil industry to enhance the recovery of heavy crude oils. Laboratory studies done by researchers in the oil industry have always shown increases in recovery with increases in temperature. Early researchers (1) attributed all of the increased recovery to the expansion of the oil at higher temperatures, but others (2,3) have found greater increases in oil recovery as the temperature increased than could be accounted for by oil expansion alone. These authors found that the more viscous oils generally showed the greatest improvement in recovery as the temperature increased. Thus, viscosity reduction appears to be a significant mechanism for improving oil recovery at higher temperatures.

The benefits of hot water for oil recovery, however, appear to be more than can be explained by oil expansion and viscosity (or viscosity ratio) reduction. Many researchers (2-5) have found changes in the permeability ratio for oil and water as the temperature increases which may contribute to an increase in oil recovery. These changes in permeability may be attributed to changes in the capillary forces of the system. An increase in temperature appears to reduce the capillary forces and allow greater oil recovery.

A field test of the hot water injection process for tertiary oil recovery demonstrated the benefits of the hot waterflooding (6). The reservoir had been previously flooded out using water at the ambient temperature of 15°C. Hot water at 190°C was then injected. In spite of severe channeling in the reservoir as indicated by temperature gradients and tracer tests, additional oil was produced from the reservoir. This additional oil, however, was produced at very high water to oil ratios, which suggests that the mechanism for oil displacement was not a more efficient displacement as would be expected by an increase in temperature, but a factor such as rate, pattern, rotation, or plugging, etc. At the end of the hot waterflood, it was found that additional cold water injection scavenged the heat remaining in the reservoir and produced additional oil.

The laboratory experiments performed here have demonstrated that moderately hot water can also be used to increase the recovery of oily wastes. The intent of these experiments was to test the proposal of using waste heat at industrial sites, such as refineries, to enhance the recovery of hydrocarbons that are contaminating the subsurface. The use of waste heat will greatly reduce the cost of the remediation process, but limitations due to the amount of waste heat available and the heat exchanging process that would be necessary is expected to restrict the maximum temperature that can be achieved for the injection water. Based on this consideration, a maximum temperature of 50°C was selected for these experiments.

Two oils have been used for the experiments. The first oil was a hydrocarbon oil used for pump lubrication. This oil likely has physical properties similar to the automatic transmission fluid and lubricating oil found contaminating the subsurface at certain sites. The second oil used was creosote, which is commonly used as a wood preservative. Creosote consists predominantly of semivolatile polynuclear aromatic hydrocarbons and phenolic compounds. It is estimated that there are 700 sites throughout the United States where wood preservation is, or has been, conducted (7), and all of these sites are likely to have some contamination due to spills or disposal practices. Currently there are more than 40 wood treatment sites on the USEPA Superfund List (8).

Experimental Methods and Materials

One dimensional displacement experiments have been performed under conditions of both constant and transient temperature to determine the effects of temperature on oil recovery. The constant temperature experiments were performed at temperatures in the range of 10°C to 50°C. A detailed description of the experimental methods was given previously (9). Stainless steel columns 7.5 cm in diameter and 45.7 cm long were used for all the experiments. The experimental method basically consists of packing the columns with sand, then flooding with carbon dioxide. Water is then introduced into the column which dissolves the carbon dioxide and allows for a complete saturation of the pore space with water. Approximately 4 pore volumes of water were allowed to flow through the column to ensure complete dissolution and removal of the carbon dioxide. Next, the column was placed in a constant temperature incubator and allowed to equilibrate to the desired temperature. Oil at the same temperature was injected by means of a constant flowrate pump to displace the water. Thus, these experiments were performed with a residual water saturation in the media.

For the constant temperature displacements, water at the same temperature as the column was injected by means of the constant flowrate pump to displace the oil. The effluent from the column was collected in a fraction collector, and the proportion of water and oil in the effluent was determined. Throughout the displacement experiments, the pressure in the water phase was measured by pressure transducers at each end of the column.

For the transient temperature displacement experiments, the column was placed in the incubator at 10°C, and water at 50°C was used to displace the oil. Again, the column effluent was collected in the fraction collector, and the pressure at each end of the column was measured. In addition, the temperature was measured using thermocouples at four locations along the length of the column during the displacement.

A constant flow rate of approximately 17 to 18 ml per minute was used for these experiments. The effect of flow rate and soil core length on oil recovery during a waterflood has been investigated (10,11), and it was found that at low flow rates, oil recovery was dependent on the flow rate and the length of the core. However, when the flow rate and the core length were increased so that the product of the core length (L), the flow rate (V), and the water viscosity (μ_w), exceeded a critical value, oil production during a waterflood became independent of flow rate and core length. The waterflood was then said to be "stabilized". The critical value of $LV\mu_w$ needed to produce a stabilized flood was somewhat dependent on the permeability of the core, but generally a value between 0.5 and 3.5 cp cm^2/min was found to be adequate (10). For the hot waterfloods performed here, the product $LV\mu_w$ ranges from 22.98 for the 10°C displacements to 9.62 cp cm^2/min for the 50°C displacements. Thus, all of the waterfloods performed here should be stabilized. All displacements were continued until at least 10 pore volumes of water had been injected, and recoveries for the various displacement experiments are compared in terms of the oil recovery at breakthrough (when water is first recovered in the column effluent) and the oil recovery after the injection of 10 pore volumes of water.

Experiments have been performed using two different oils and two different sands. The first oil used was a hydrocarbon oil which is 100 percent neutral paraffins

(Inland 15) and is used for vacuum pump lubrication. This is a light oil, with density changing approximately linearly with temperature in the range of 10°C to 50°C, ranging from 0.892 to 0.8472 gm/cm³. The second oil used was creosote. The creosote was first "weathered" in the laboratory to help stabilize its physical properties. This was done by mixing the creosote with water in a separatory funnel to remove the water soluble constituents, and then exposing the creosote to the atmosphere in a hood to allow the volatile components to escape. These same processes of solubilization and volatilization would occur for a spill of creosote that is exposed to air and water. The physical properties of this "weathered" creosote were then measured. The density of the creosote also varied approximately linearly with temperature between 10°C and 50°C, ranging from 1.1069 to 1.0784 gm/cm³. This density is high compared to the average values for creosote, which indicates that most of the lighter components have probably been lost due to volatilization or solubilization.

The ratio of the oil viscosity to water viscosity as a function of temperature is shown in Table I. Both of the oils show an exponential decrease in viscosity as the temperature is increased, which is the expected relationship between the viscosity of a liquid and temperature. Attempts were made to determine the amount of the oils that were volatile at different temperatures. Open beakers of Inland 15 and creosote were placed in a constant temperature water bath, and held for 20 days at temperatures as high as 30°C. Less than 1 weight percent of the Inland 15 was lost during this time period, while 7.5 weight percent of the creosote was lost. Thus, it would be expected that only negligible amounts of the Inland 15 would be volatilized during these displacement experiments. However, this may not be the case for the creosote. In a separate experiment, a creosote sample of known weight was placed in an open beaker contained in a water bath and the weight was determined every 24 hours. For the first 24 hours, the water bath was at 20°C, and approximately 0.5 weight percent of the creosote was lost. For the next 24 hours, the same sample was held at 30°C, and an additional 1.4 weight percent was lost due to volatilization. During the next 24 hours at 40°C, an additional 2.7 percent of the creosote was volatilized, and 24 hours at 50°C volatilized another 4.9 percent. Thus, significant creosote volatilization occurs at temperatures above 30°C. However, it is not possible to determine how much of the creosote was volatilized as it was displaced from the soil column and collected in the fraction collector.

The surface tension of the oils and the interfacial tension of the oils with water were measured using a du Nouy Ring Tensiometer (ASTM method D-971), and the results are shown in Table II. Each tension given is the average of at least three measurements. Note that although the two oils have similar surface tensions, the interfacial tension of creosote with water is significantly lower than that for Inland 15. The average tensions at different temperatures were compared using Student's t-test.

Table I. Ratio of Oil Viscosity to Water Viscosity

	10°C	20°C	30°C	40°C	50°C
Inland 15/Water	85.5	58.6	44.4	35.8	28.1
Creosote/Water	27.3	19.8	15.5	13.1	11.3

Table II. Surface and Interfacial Tensions (dynes/cm)

	10°C	20°C	30°C	40°C	50°C
Inland 15/Air	29.22 ±0.40	30.34 ±0.65	30.18 ±0.44	26.37 ±0.31	25.25 ±0.22
Inland 15/Water	39.96 ±1.85	40.03 ±0.34	40.03 ±0.34	39.70 ±0.88	39.99 ±0.32
Creosote/Air	32.40 ±1.52	33.47 ±0.28	28.98 ±2.06	25.02 ±0.84	26.83 ±1.50
Creosote/Water	9.65 ±1.98	7.83 ±2.12	6.16 ±1.87	5.31 ±2.03	5.90 ±1.48

Although the surface tension of Inland 15 is essentially the same in the temperature range of 10° to 30°C, it is significantly lower at the 98 percent confidence level at 40° and 50°C. The interfacial tension of the Inland 15 with water did not change with temperature in the range of temperatures used here. At the 98 percent confidence level the surface tension of the weathered creosote decreased with temperature as the temperature was increased past 30°C. The interfacial tension of the creosote appears to decrease slightly as the temperature increases, but these changes are not statistically significant at the 98 percent confidence level.

Based on previous research on the effect of temperature on the interfacial tension of oils, it is difficult to predict what the effects of temperature should be. Some oils have fairly constant decreases in interfacial tension as the temperature increases, but other oils have been found to have different temperature relationships (*4,12*). Although the oils did generally show a decrease in interfacial tension with temperature increase, in the lower temperature ranges the interfacial tension sometimes remained constant with temperature or showed a slight increase. Thus, the results found here are not inconsistent with the results of other researchers.

Both sands used were silica sands which differed only in their distribution of grain sizes. The first, referred to as 20/30 sand, has a very uniform grain size in the range of 0.85 to 0.60 mm. The second, called mixed sand, contained an equal amount of grains within three different size ranges: 0.85 to 0.60 mm, 0.60 to 0.25 mm, and 0.25 to 0.15 mm. The measured intrinsic permeability of the 20/30 sand at 10°C was 2.07×10^{-6} cm^2 ± 4.15×10^{-6}. A small decrease in intrinsic permeability with temperature was noted, and at 50°C the measured permeability was 5.29×10^{-7} cm^2 ± 5.57×10^{-7}. For the mixed sand, there were not enough measurements of permeability to determine if it changed with temperature. The average measured intrinsic permeability was 5.26×10^{-7} cm^2 ± 3.40×10^{-7}. Other researchers (*13,14*) have measured decreases in permeability with temperature increase.

Experimental Results

Constant Temperature Displacements. The initial oil saturation, pore volumes of oil recovered at breakthrough, and pore volumes of oil recovered after the injection of

Table IIIa. Oil recovery data for the Inland 15 oil and 20/30 sand

	10°C	20°C	30°C	40°C	50°C
Initial Oil Saturation	1.000	0.998	0.984	0.989	0.978
Pore Volumes Recovered at Breakthrough	0.248	0.272	0.287	0.288	0.323
Pore Volumes Recovered at 10 Pore Volumes	0.613	0.637	0.647	0.704	0.689
Recovery Between Breakthrough and 10 Pore Volumes	0.365	0.365	0.361	0.416	0.366
Final Oil Saturation	0.391	0.384	0.344	0.295	0.299

Table IIIb. Oil recovery data for the Inland 15 oil and mixed sand

	10°C	30°C	50°C
Initial Oil Saturation	0.982	0.983	0.975
Pore Volumes Recovered at Breakthrough	0.249	0.334	0.342
Pore Volumes Recovered at 10 Pore Volumes	0.657	0.698	0.724
Recovery Between Breakthrough and 10 Pore Volumes	0.408	0.364	0.382
Final Oil Saturation	0.331	0.291	0.232

10 pore volumes of water are given in Table III. The table shows that for Inland 15, as the temperature was increased, the breakthrough of water at the effluent end of the column was delayed, therefore, more oil was recovered before breakthrough. This is a result of the lower oil viscosity at higher temperatures which increases the mobility of the oil (15). This was where most of the increased oil recovery at the higher temperatures was realized for the Inland 15, as the percent recovery between

Table IIIc. Oil recovery data for creosote in 20/30 sand

	10°C	20°C	30°C	40°C
Initial Oil Saturation	1.000	1.000	0.964	0.978
Pore Volumes Recovered at Breakthrough	0.316	0.338	0.332	0.310
Pore Volumes Recovered at 10 Pore Volumes	0.630	0.674	0.700	0.677
Recovery Between Breakthrough and 10 Pore Volumes	0.313	0.335	0.367	0.368
Final Oil Saturation	0.370	0.326	0.264	0.296

breakthrough and 10 pore volumes of water injection remained essentially constant with temperature, particularly for the 20/30 sand.

Table III also shows that the mixed sand had later breakthroughs and overall greater recoveries than the 20/30 sand. Intuitively, it would seem that the uniform pores of the uniform grain size sand would more closely approach plug flow, and therefore more of the oil would be displaced from the 20/30 sand. However, calculations made using the Buckley and Leverett equation (15) for two phase flow and the properties of the two sands and the Inland 15 show that greater recoveries, both at breakthrough and after the injection of 10 pore volumes of water, may be expected from the mixed sand due to its wider range of pore sizes (9). Most of the laboratory research that has been done does not address the question of the effect of the grain size or pore size distribution on the amount of oil that can be recovered. Experiments with oil-wet sands show an increase in breakthrough recovery as the permeability and porosity increased, but ultimate recovery was found to be inversely related to the permeability (16). Also, the effects of the properties of the cores on oil recovery was not as great as the effects of wettability or interfacial tension. More research is needed before conclusions can be drawn on the effects of the properties of the soil on oil recovery.

Table IIIc shows the results of constant temperature hot waterfloods using creosote in the 20/30 sand. The creosote had approximately the same breakthrough at all temperatures. At 10°C, essentially all of the creosote that was recovered came during the first 5 pore volumes of water injection. As the temperature was increased to 20° and 30°C, increasing amounts of creosote were recovered during the latter portion of the waterflood. The displacement at 40°C, however, recovered somewhat less oil than the 30°C waterflood. Thus the amount of oil recovered between breakthrough and 10 pore volumes of throughput is dependent on temperature and increases as the temperature increases, at least until a certain temperature is reached.

The experiments done to determine the percent volatiles in the creosote indicate that at 40°C, the percent volatiles may be high enough to affect the oil recovery results. Thus, the oil recoveries shown for creosote are likely lower than what was actually recovered. Because there is no way to determine just how much of the creosote is lost due to volatilization during the experiments, the oil recovery results cannot be adjusted accordingly. Thus, the measured oil recoveries for creosote are likely less than the total amount of creosote extracted from the column. This would mean that the calculated amounts of creosote remaining in the column are high.

These recoveries can be compared at least in a qualitative way with the recoveries of other researchers. Early researchers (1) found 2 to 15 percent increases in recoveries of oil when the temperature was raised from 26 to 165°C, while displacement experiments with other oils (2) showed a 12 to 20 percent increases in oil recovery for temperature increases from 24 to 149°C (these percent recoveries are based on the initial oil volume in the column). Comparing the recoveries achieved here to those recoveries which were achieved with much larger increases in temperature shows that these results are generally better than those found by researchers in the oil industry. Recoveries of 60 to 70 percent for coal gasification wastes and 84 to 94 percent for wood treating wastes (17) are about the same or better than the recoveries achieved here by the injection of 10 pore volumes of water, but information on the temperatures used and increased recoveries achieved by the use of hot water was not provided.

Recent research at field sites has shown that when hydrocarbons are spilled to the subsurface, they may reside in the pore space at saturations significantly less than one. Diesel fuel in unconsolidated sands was found at saturations ranging from 0.05 to 0.50 (18), while the saturation of crude oil from a pipeline spill ranged from 0.15 to as high as about 0.85 (19). Displacements when the initial oil saturations are significantly less than one will follow the same pattern of oil saturation behind the water front, but would start at the saturation corresponding to the intial saturation at the site. The initial saturation will not effect the saturation found behind the flood front, just the amount of oil that is actually recovered. Thus, the percent oil recoveries are dependent on the initial oil saturations, but the final oil saturations, and the percent reductions in remaining oil saturation, are not.

Comparison of Table IIIa and Table IIIc shows that breakthrough for the displacement of creosote is always later than the displacement of Inland 15 at the same temperature. Also, there is always considerably more creosote recovered during the 10 pore volumes of water injection. Thus, lower oil saturations were found at each temperature for the creosote than for the Inland 15, with the exception of the 40°C waterfloods which did surprisingly well at recovering Inland 15 and surprisingly poor at recovering creosote.

The differences in recovery for the two oils is likely a reflection of the differences in their interfacial tensions and wettabilities. Studies on the displacement of a nonwetting phase by a wetting phase (i.e., oil being displaced from a water wet media), showed very little oil recovered after breakthrough (less than 0.1 pore volume). In this case, decreases in the interfacial tension cause only small increases in the oil recovery. For the displacement of a wetting phase by a nonwetting phase (i.e., displacement of oil from oil-wet media), breakthrough of water is earlier, and significant quantities of oil are produced after breakthrough. In the case of an oil-wet media, significant increases in oil recovery were found as the interfacial tension decreased (*16,20*).

Comparing the oil recovery curves found here with those found for oil-wet media (*20*) would indicate that the silica sand used here is not strongly water-wet nor strongly oil-wet in the presence of Inland 15. The constant pore volumes of Inland 15 recovered between breakthrough and 10 pore volumes of water injection would indicate that the wettability of this system does not change with temperature. For the creosote, the recovery curves also appear to be of some intermediate wettability, but there also appears to be a trend toward greater oil-wetness as the temperature increases. This statement is made based on the increases in oil recovery after breakthrough which was found as the temperature increased. The creosote/water/silica sand system appears in general to be more water-wet than the system with Inland 15.

Some researchers (*21,22*) have found that emulsions are formed when hot water or steam are used to displace oil from porous media. The formation of emulsions would adversely affect efforts to recycle the water and would likely increase the amount of treatment that the water would require before it could be discharged. Visual inspection of the column effluent from these experiments showed that emulsions were not formed.

The density of the creosote in the effluent of the column was measured at the conclusion of two displacement experiments to give an indication of whether or not the creosote leaving the column had the same composition as the creosote that was pumped into the column. Measurements were made after the completion of both a 10°C and a 40°C displacement, and these results show that the density of the creosote leaving the column was essentially the same as the density of the creosote that had entered the column. Thus it appears that there was not a significant chromatographic separation of the creosote in the column during the displacement. This lack of a change in oil density may indicate also that the amount of volatiles lost from the column effluent was not significant, although the odor was significant.

Figure 1 shows the pressure drop in the column during the displacement of Inland 15 from the mixed sand at temperatures of 10, 30, and 50°C. Graphs of the pressure drop for Inland 15 and creosote displacement from 20/30 sand have the same shape, but are lower in magnitude. The maximum pressure drop in the water phase occurs at breakthrough, when the water first begins to flow from the column. After breakthrough, the pressure drop along the column decreases at a rate that appears to be related to the rate of displacement of oil from the column, which in turn is related to the rate of increase of pore space in the column in which the water is flowing. For most of the displacement experiments, the pressure drop reached its approximately constant value after the injection of about 2 pore volumes of water. (The time scale of this graph can be converted to approximate pore volumes by taking into account that at the flow rate used, it took about 35 minutes for the injection of one pore volume of water.) The exception to this is the 50°C curve shown here, which shows that the pressure drop did not reach its constant value until about 6 pore volumes of water had been injected.

The pressure drops measured during these displacement experiments correspond to maximum hydraulic gradients during the displacements of 2 to 6.6 for the Inland 15, and 0.94 to 0.46 for the creosote. The gradients used here for the Inland 15 probably cannot be achieved in the field. However, the use of smaller gradients and

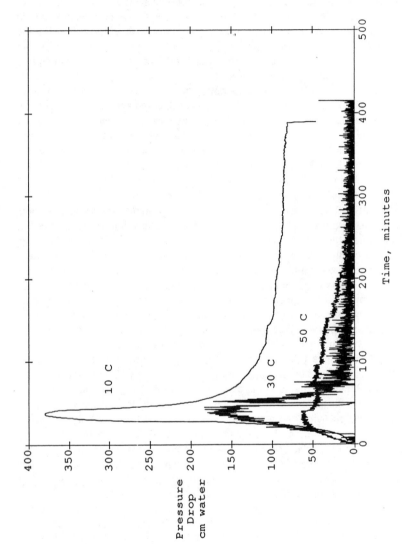

Figure 1. Pressure drop during the displacement of Inland 15 from mixed sand at 10, 30, and 50°C.

therefore smaller velocities in the field should not greatly affect the displacement results in a homogeneous porous media, as the greater distances in the field over which the displacement occurs should still produce a stabilized flood (*10*). However, it has been found theoretically (*15*) and under field conditions (*6,23*) that in a heterogeneous system, higher flow rates may improve the sweep efficiency of a waterflood, which will in turn increase the oil recovery.

A mass balance calculation on the amount of oil originally contained in the column and the amount recovered was used to calculate the residual oil remaining in the column after the injection of 10 pore volumes, and the results are shown in Table III. It should be noted that for some of the displacements, oil could still be recovered after the injection of 10 pore volumes of water, although at high water to oil ratios. Thus, these are not true residual or "irreducible" oil saturations. The graph shows that raising the temperature from 10°C to 50°C in a system containing Inland 15 decreases the residual oil in the column after the injection of 10 pore volumes from 39 percent to 30 percent for the 20/30 sand, and from 33 percent to 25.5 percent in the mixed sand. This is a 25 to 33 percent reduction in the amount of residual oil held in the pore spaces that was achieved using a moderate increase in the temperature of the flood water. For the creosote, the residual oil saturation decreased from 37 percent at 10°C to 26.4 percent at 30°C, which is a 29 percent decrease in residual creosote. The lower oil recovery between breakthrough and 10 pore volumes of water injected at 40°C left a greater residual oil at this temperature than at 30°C, but it was still 20 percent lower than the residual at 10°C. Thus, both oils and both sands show significant reductions in residual saturations for moderate temperature increases.

It is interesting to note that for the 20/30 sand, the recovery of Inland 15 was greater at 40°C than at 50°C, and the recovery of creosote was greater at 30°C than at 40°C. This may indicate that there is an optimum temperature at which the displacement is most efficient, and that this optimum temperature is dependent on the oil. It might be expected that these displacements would show an optimum based on the fact that the decrease in viscosity ratio with temperature increase drops off quickly in this range of temperatures. However, existence of an optimum temperature for the displacement cannot be confirmed without additional data.

Transient Temperature Displacements.   The oil recoveries achieved during the transient temperature experiments using the Inland 15 oil are shown in Table IV for both the 20/30 and mixed sands. These experiments more closely reproduce what would happen in the field where the soil and oil are at ambient temperature, and water at a higher temperature is injected to displace the oil. Comparison of Tables IIIa and IIIb with Table IV shows that breakthrough was not significantly delayed from what it had been at 10°C, but as the higher temperature water moves through the column, raising the temperature as it goes, the oil recovery increases significantly over that can be achieved with a 10°C waterflood. The final oil saturation for both sands after the injection of 10 pore volumes of water was essentially the same as the residual oil saturation found in a 40°C constant temperature waterflood.

Temperature measurements along the column show that the heat loss during the displacement is fairly significant, limiting the high temperature in the column to less than 40°C, and dropping off fairly quickly and uniformly along the column to a low temperature of around 30°C near the effluent end. The measurements seem to indicate

Table IV. Oil recovery data for the transient temperature displacements

	20/30 sand	Mixed sand
Initial Oil Saturation	0.987	0.994
Fraction (wt.) Recovered at Breakthrough	0.261	0.270
Fraction (wt.) Recovered at 10 Pore Volumes	0.654	0.704
Recovery Between Breakthrough and 10 Pore Volumes	0.393	0.435
Final Oil Saturation	0.297	0.255

that very little heat travels in front of the hot water bank, but the temperature at any place along the column reaches its equilibrium temperature fairly quickly once the hot water front reaches it. Undoubtedly, the stainless steel column used for these experiments moved heat away from the sand faster than would occur in a field situation because of the significantly higher thermal diffusivity of stainless steel than sand. Thus, actual temperatures in the field may be significantly higher, and consequently the oil recovery may be greater.

The transient temperature experiments had maximum pressures that were essentially the same as those found in the 10°C constant temperature experiments, and during the latter stages of the experiment the pressure level dropped off to that of the 30°C to 40°C displacements. This can be expected due to the fact that the temperature logs for these experiments show that essentially no heat was transported in front of the hot water front. Thus, the oil bank being displaced ahead of the hot water front had the viscosity of 10°C oil. After breakthrough, temperatures in the column ranged from approximately 40°C at the influent to 32°C at the effluent, and the pressure of the system was similar to that at the end of a 30°C waterflood.

Conclusions

Constant temperature hot waterfloods using Inland 15 pump oil and creosote have shown that moderately hot water can significantly increase the amount of oil that is recovered from sands. Residual saturations in the column after the injection of 10 pore volumes of water were reduced by 25 to 33 percent. The two oils were found to have different responses in their oil recovery curves to the increase in temperature. Most of the increase in recovery for the Inland 15 oil came at the early portion of the waterflood through a delay in breakthrough, and this is likely due to the significant decrease in the viscosity of the oil as the temperature increases. For creosote, breakthrough was essentially independent of temperature and the additional recovery at higher temperatures came during the later stages of the hot waterflood. This response to the temperature increase is likely due to the fact that only small decreases in viscosity are achieved over this temperature range. The interfacial tension also shows a small decrease over this range of temperatures, and this could account for the difference in behavior after breakthrough.

Although these constant temperature experiments provide useful information on the effects of temperature on recovery and the two phase flow properties of the oil, they do not represent the case of what would happen in the field. The purpose of the transient temperature experiments was to better represent the field conditions, and these experiments have shown that the benefits of hot water can be achieved under conditions similar to what would occur in the field.

Although the hot waterfloods produced a significant reduction in the amount of oil remaining in the pores after the injection of 10 pore volumes, the remaining oil saturation is likely to be above cleanup goals. Thus, it is likely that additional treatment, such as biodegradation, will be required. However, this substantial reduction in the amount of residual saturation should significantly reduce the time and expense required for bioremediation, and the overall time for cleanup should be reduced by the use of a hot waterflood as a first step in the remediation process.

The hot waterflooding process for the recovery of oily contaminants is generally applicable to viscous oils which are present in the pore space in quantities greater than their residual saturation. A field trial of hot waterflooding done in an oil reservoir (6) demonstrated that hot water may also improve the area that is swept by water in a heterogeneous system, and thereby also increase water recovery. The use of waste heat for the energy input will greatly assist in keeping the cost of the remediation process at reasonable levels. Advantages of the hot water process over processes such as surfactant flooding and cosolvent flooding include the fact that the hot water process does not require the injection of potentially harmful chemicals to the subsurface in order to recover the contaminants. When compared to remediation techniques such as steam injection, the advantages of the hot water process include the fact the significantly lower temperatures and pressures are easier to produce and handle in the injection equipment and wells, etc. Although the temperatures used for steam injection are high enough to be very detrimental to the microorganisms commonly found in the subsurface, the moderate temperatures used here may be beneficial to subsequent bioremediation.

Acknowledgments

Although the research described in this article has been supported by the U. S. Environmental Protection Agency, it has not been subjected to the Agency review and therefore does not necessarily reflect the views of the Agency.

Literature Cited

1. Willman, B. T.; Valleroy, V. V.; Runberg, G. W.; Cornelius, A. L.; Powers, L. W. *J. Pet. Tech.*, **1961**, 681-690.
2. Edmondson, T. A. *J. Can. Pet. Tech.*, **1965**, 236-242.
3. Davidson, L. B. *J. Pet. Tech.*, **1969**, 1037-1046.
4. Poston, S. W.; Ysrael, S.; Hossain, A. K. M. S.; Montgomery III, E. F.; Ramey, Jr., H. J. *Soc. Pet. Eng. J.*, **1970**, 171-180.
5. Quettier, L.; Corre, B. *SPE Reservoir Eng.*, **1988**, 149-157.

6. Martin, W. L.; Dew, J. N.; Powers, M. L.; Steves, H. B. *J. Pet. Tech.*, **1968,** 739-751.

7. Mueller, J. G.; Chapman, P. J.; Pritchard, P. H. *Environ. Sci. Technol.*, **1989,** *23,* 1197-1201.

8. Rosenfeld, J. K.; Plumb, Jr., R. H. *Ground Water Mon. Rev.,* **1991,** 133-140.

9. Davis, E. L.; Lien, B. K. *Laboratory study on the use of hot water to recover light oily wastes from sands;* EPA/600/R-93/021; U. S. EPA, R. S. Kerr Environmental Research Laboratory, Ada, OK, 1993.

10. Rapoport, L. A.; Leas, W. J. *Trans. AIME,* **1953,** *198,* 139-148.

11. Perkins, Jr., F. M. *Trans. AIME,* **1957,** *210,* 409-413.

12. Lo, H. Y.; Mungan, N. *Soc. Pet. Eng. J.,* **1973,** 1-12.

13. Weinbrandt, R. M.; Ramey, Jr., H. J.; Casse, F. J. *Soc. Pet. Eng. J.,* **1975,** 376-384.

14. Casse, F. J.; Ramey, Jr., H. J. *J. Pet. Tech.,* **1979,** 1051-1059.

15. Buckley, S. E.; Leverett, M. C. *Trans. AIME,* **1942,** *146,* 107-116.

16. Warren, J. E.; Calhoun, Jr., J. C. *Trans. AIME,* **1955,** *204,* 22-29.

17. Johnson, L. A. report submitted for the DNAPL Meeting, U. S. EPA, R. S. Kerr Environmental Research Laboratory, Dallas, TX, April 17-18, 1991.

18. Huntley, D.; Hawk, R. N.; Corley, H. P. *Ground Water,* **1994,** *32,* 626-634.

19. Hess, K. M.; Herkelrath, W. N.; Essaid, H. I. *J. Contam. Hydrol.,* **1992,** *10,* 75-96.

20. Mungan, N. *Soc. Pet. Eng. J.,* **1966,** 247-253.

21. Stewart, L. D.; Udell, K. S. *SPE Reservoir Eng.,* **1988,** 1233-1242.

22. Bennion, R. G.; Thomas, F. B. *Pet. Soc. of CIM,* **1983,** 1-17.

23. Borisov, Y. P.; Babalyan, G. A.; Ivanova, M. M.; Zheltov, Y. P.; Orduzhev, S. A.; Rosenberg, M. D.; Satarov, M. M.; Krilov, A. P. *Eighth World Petroleum Congress;* 1971; Vol. 3, pp 299-306.

RECEIVED March 14, 1995

BIOLOGICAL   DEGRADATION

Chapter 20

Metabolism of Alkanes by *Rhodococcus erythropolis*

John Lofgren, Sandra Haddad, and Kevin Kendall

Department of Cell and Molecular Biology, Tulane University, New Orleans, LA 70118

Twenty one mutants of *R. erythropolis* were isolated that were unable to use tridecane and/or tetradecane as carbon sources. The mutants were grouped into four general classes based upon their alkane utilization profiles; Class I mutants were unable to grow on any alkanes. Class II mutants could grow only on alkanes of 14 carbon atoms or longer, whereas Class III mutants could grow only on alkanes of 13 carbon atoms or less. The Class IV mutants were able to grow only on alkanes containing an odd number of carbon atoms. No enzymatic defects related to alkane metabolism were found in any of the mutants. Four of the seven Class I mutants were incapable of growth using even chain-length fatty acids. It appears that these mutants, and the three Class IV mutants, may be defective in the glyoxylate bypass across the tricarboxylic acid cycle.

Although naturally occurring microorganisms offer tremendous potential for the bioremediation of toxic waste sites, genetically engineered microorganisms may eventually provide a much more efficient means of achieving this aim. In order to create genetically manipulated microorganisms, the metabolic pathways responsible for the biodegradation of toxic compounds have to be understood and the genes encoding the appropriate enzymes must be cloned and analyzed in detail. Currently, the best genetically understood metabolic pathways are the toluene biodegradation pathway of *Pseudomonas putida* (*1*) and the alkane metabolic pathway of *Pseudomonas oleovorans* (*2*). Although many other metabolic pathways are now being studied, the majority of these are found in other species of *Pseudomonas* (*3, 4*). However, there are other groups of bacteria that have potential for development as bioremediation organisms. Foremost amongst these are the Actinomycetes, a diverse family of sapprophytic soil bacteria that have recently been shown to be capable of degrading a wide variety of recalcitrant compounds. In particular, the 'nocardioform' group, consisting of the genera *Nocardia*, *Mycobacterium* and *Rhodococcus*, have been reported to degrade compounds ranging from aliphatic (*5*) and aromatic (*6*) hydrocarbons to dioxane (*7*), pyridine (*8*), nitrile compounds (*9*), aniline derivatives (*10*), and even chlorinated compounds such as pentachlorophenol (*11*) and trichloroethene (*12*). However, despite the abundance of reports on the ability of Actinomycetes to degrade these problematic compounds, there is very little known of

0097–6156/95/0607–0252$12.00/0

the basic genetics and biochemistry of the metabolic pathways used to degrade even the simplest of these compounds.

We have been investigating the genetics and biochemistry of hydrocarbon metabolism by Actinomycetes with a view to eventually creating genetically manipulated strains with enhanced bioremediation potential. This report focusses upon the metabolism of alkanes by *Rhodococcus erythropolis* (ATCC 4277). From preliminary biochemical analysis, we show that the strain may use a sub-terminal oxidation pathway to metabolize *n*-alkanes. We also describe the isolation and characterization of 21 mutants of the strain that can no longer utilize either tridecane and/or tetradecane as a sole source of carbon and energy.

Materials and Methods

Bacterial Strains and Growth Conditions. The bacterial strains used in this study (Table 1) were obtained directly from the American Type Culture Collection (ATCC) and were maintained on LBA at 30°C. Carbon source utilization experiments used RMM (0.5g/l K_2HPO_4, 2g/l $(NH_4)_2HPO_4$, 2 ml/l trace elements [*13*], 0.2g/l $MgSO_4$, 1.5% Difco 'Bacto' agar) adjusted to pH 7.2 with HCl. Carbon sources were generally supplied at 0.4% final concentration unless otherwise specified.

Substrate Growth Tests. Tests were performed on RMM supplemented with an appropriate carbon source. Soluble non-volatile substrates, including sucrose, glucose and most fatty acids, were added into RMM at 0.4% (w/v) before the plates were poured. Volatile hydrocarbons were supplied by pipetting 100μl onto filter disks soaked with water placed in the lids of petri-dishes which were then sealed with parafilm. All plates were incubated at 30°C for 3-5 days. Growth on hydrocarbons was scored in comparison to growth on RMM supplemented with a preferred non-hydrocarbon such as glucose or sucrose. The growth of *R. erythropolis* ATCC 4277 on hydrocarbons was always compared relative to growth on its preferred carbon source, sucrose.

UV Mutagenesis. *R. erythropolis* was grown in LB at 30°C to an A_{600} of 0.2 - 0.4. A 20 ml sample was briefly sonnicated to break up aggregated cells then placed into a petri dish. The dish was placed under a germicidal UV lamp (at a distance of 10 cm) and agitated with a stirbar for 12.5 minutes. Mutagenized samples were then grown in the dark for at least two generation times (5 hours) before plating on LBA. Approximately 99% of the cells were killed by this treatment. After 3-4 days incubation at 30°C on LBA, the resulting colonies were replica-plated onto RMM containing sucrose, tridecane or tetradecane as the sole source of carbon and energy. Mutants that grew normally on sucrose, but failed to grow on either tridecane and/or tetradecane were analyzed furthur.

Preparation of Cell Free Extracts. Strains were grown at 30°C in LBA for 4 days, then harvested and resuspended in liquid RMM (i.e. without agar) supplemented with 0.4% tetradecane. After 48 hours additional growth at 30°C, the cells were harvested and resuspended in approximately 3ml per gram wet weight in 20mM Tris.Cl pH 7.0. "Bead Beater" (Biospec products, Bartlesville, OK.) lysis was achieved with either 10 x 20 second bursts in a micro chamber or 8 x 45 second bursts in the standard chamber. Cell debris and unlysed cells were removed by centrifugation at 2,000g for 5 minutes. A cleared lysate was obtained after centrifugation at 30,000g for 30 minutes to remove membranes and other insoluble material. The protein concentration in cell extracts was assayed by the method of Bradford (*14*) using a commercial kit from Bio-Rad (Richmond, CA.).

Enzyme Assays. Primary and secondary alcohol dehydrogenase activities were assayed by following the alcohol-dependent reduction of NAD^+ by cell extracts. 5-100µl of extract was incubated at 37°C with 50mM Tris.Cl pH9.0, 10mM $MgCl_2$, 2.5mM NAD^+ and 0.1% alcohol in a final volume of 1ml. 1-octanol and 2-octanol were used to detect the primary and secondary alcohol dehydrogenases respectively. The formation of NADH was followed at A_{340} by a Milton Roy 3000 Array spectrophotometer equipped with kinetics software. Activities (Units) were expressed as micromoles of NADH formed per minute at 37°C. Specific activities were generally expressed as milliUnits (mU) per mg of protein. Aldehyde dehydrogenase was assayed in a similar manner except that octanal was supplied at a concentration of only 0.025% in the assay mix.

Alkyl acetate esterase was detected indirectly as follows; 20-100µl of cell extract was added to a reaction mixture containing 50mM Tris.Cl pH 9.0, 10mM $MgCl_2$, 2.5mM NAD^+ and 0.1% octyl acetate in a total volume of 1ml. NAD^+ reduction to NADH was followed at A_{340}. After an initial lag phase, the rate of NAD^+ reduction was found to increase slowly until a steady-state rate was achieved. This was assumed to be due to an esterase activity cleaving the octyl acetate to form 1-octanol; this would then be oxidized by the primary alcohol dehydrogenase present in the extract, causing a concommitant reduction of NAD^+. Control experiments showed that the increase in A_{340} was absolutely dependent upon the addition of both cell extract and octyl acetate. In addition, no NAD^+ reduction was found when purified primary or secondary alcohol dehydrogenases from *R. erythropolis* were incubated with octyl acetate (data not shown). Conversely, addition of purified primary alcohol dehydrogenase to cell extracts caused a decrease in the lag phase of the octyl acetate reaction and gave slightly higher rates of NAD^+ reduction. All of these observations are consistent with an alkyl acetate esterase releasing 1-octanol from octyl acetate and so permitting the detection and quantification of primary alcohol dehydrogenase activity. Alkyl acetate esterase activities were thus expressed as 'minimal' milliUnits of activity to reflect the semi-quantitative nature of the assay.

Results

Hydrocarbon Utilization by Actinomycetes. 34 strains of Actinomycetes representing 9 genera were obtained from the ATCC and screened for their ability to utilize a range of hydrocarbons as sole sources of carbon and energy. A summary of the results obtained for 29 of these strains (belonging to six genera), including the well characterized species *Streptomyces lividans* (*15*), is presented in Table I. Representatives of three additional genera, *Brevibacterium divariticum* (ATCC 21792), *Curtobacterium pusillum* (ATCC 19096) and *Cellulomonas fimi* (ATCC 8183) failed to show significant growth on any of the compounds tested. Table I shows that the ability to use *n*-alkanes as sole sources of carbon and energy is widespread within the Actinomycetes, with representatives of *Arthrobacter*, *Corynebacterium*, *Mycobacterium*, *Nocardia*, *Rhodococcus* and *Streptomyces* capable of utilizing alkanes ranging in length from dodecane to hexadecane. Few strains were capable of utilizing decane, and no strains showed significant growth on hexane under the conditions used (100µl of the volatile substrate per 25ml agar plate). However, additional experiments (see below), have demonstrated that at least one strain, *R. erythropolis* (ATCC 4277), can grow at the expense of hexane if the hydrocarbon is supplied at a lower concentration (25µl per 25ml plate).

Aromatic hydrocarbons were less well utilized. Eight strains were able to utilize toluene, although only three of these grew well (ATCC 19070, 29678 and 15077). Similarly, xylenes, trimethyl benzene and naphthalene were also poorly utilized, although some strains were capable of growth on these compounds. Conversely, most strains were able to use benzoate as a carbon source. This is

Table I. Growth of Actinomycetes on Various Hydrocarbons [a]

Species	ATCC	no carbon source	hexadecane	pentadecane	tetradecane	tridecane	dodecane	decane	octane	naphthalene	toluene	m-xylene	p-xylene	o-xylene	trimethylbenzene	benzaldehyde	benzyl alcohol	benzoate	cyclohexane	cyclohexanone
Arthrobacter globiformis	35698	-	-	-	-	+	-	-	-	-	-	-	-	-	-	o	o	+	-	+
Arthrobacter paraffineus	15590	-	+	+	+	+	+	+	+	-	-	+	-	-	-	-	-	+	-	+
Arthrobacter petroleophagus	21494	-	+	+	+	+	+	+	+	-	+	-	+	-	+	o	+	+	-	+
Arthrobacter picolinophilus	27854	-	+	+	+	+	+	+	-	-	-	-	-	-	-	+	o	+	+	-	+
Arthrobacter sp.	21908	-	+	+	+	+	+	+	+	-	-	-	-	-	-	-	+	+	-	+
Arthrobacter sp.	33790	-	-	+	+	-	-	-	-	-	-	-	-	-	-	o	-	+	-	+
Arthrobacter sp.	43561	-	-	-	+	+	-	-	-	-	-	-	-	-	-	o	-	o	-	+
Corynebacterium petrophilum	21404	-	+	+	+	+	+	-	-	o	-	-	-	-	-	o	-	+	-	-
Corynebacterium sp.	29355	-	-	-	+	+	+	+	-	o	-	-	-	-	-	o	o	-	-	-
Mycobacterium petroleophilum	21497	-	-	-	-	-	+	-	-	o	+	-	-	-	-	o	-	-	-	-
Mycobacterium phlei	15610	-	+	+	+	+	+	-	-	-	+	-	-	-	+	o	+	+	-	+
Mycobacterium vaccae	29678	-	+	+	+	+	+	+	+	-	+	+	+	+	+	o	-	-	-	+
Nocardia globerula	19370	-	+	+	+	+	+	+	-	-	-	-	-	-	-	o	+	+	-	-
Nocardia globerula	21505	-	+	+	+	+	+	-	-	-	-	-	+	-	o	+	+	-	+
Norcardia petroleophila	15777	-	+	+	+	+	-	-	-	+	-	-	+	-	-	o	o	o	-	+
Nocardia sp.	29100	-	+	+	+	+	+	+	-	-	-	-	-	-	o	o	-	+	-	+
Nocardia sp.	21145	-	+	+	+	+	+	+	-	+	-	-	+	-	+	-	+	+	-	+
Rhodococcus erythropolis	4277	-	+	+	+	+	+	+	-	-	-	-	-	-	-	-	+	+	-	+
Rhodococcus rhodochrous	14347	-	+	+	+	+	+	-	-	-	+	+	+	+	+	-	+	+	-	+
Rhodococcus rhodochrous	21243	-	+	+	+	+	+	+	-	-	-	-	-	+	-	-	+	+	-	+
Rhodococcus sp.	19070	-	+	+	+	+	+	+	-	-	+	+	+	+	+	-	+	+	-	-
Rhodococcus sp.	21146	-	+	+	+	+	+	+	-	+	-	-	+	-	+	-	+	+	-	+
Rhodococcus sp.	21504	-	+	+	+	+	+	-	-	-	-	-	-	-	+	-	+	+	-	+
Rhodococcus sp.	21507	-	+	+	+	+	+	-	-	-	-	-	-	-	-	-	+	+	-	-
Rhodococcus sp.	29671	-	+	+	+	+	+	+	-	-	-	-	-	+	-	-	+	+	-	+
Rhodococcus sp.	29673	-	+	+	+	+	+	+	-	-	-	-	-	-	o	o	-	+	-	+
Streptomyces badius	39117	-	-	-	-	-	-	-	-	o	-	-	-	-	-	o	-	+	-	o
Streptomyces coriofaciens	14155	-	-	-	-	-	-	-	-	o	-	-	-	-	-	o	o	-	+	-	o
Streptomyces lividans	TK64	-	-	-	-	-	-	-	-	-	-	-	-	-	-	o	o	-	-	-
Streptomyces setonii	39116	-	+	+	+	+	+	-	-	-	-	-	-	-	-	o	+	+	-	-
Streptomyces sp.	15077	-	+	+	+	+	+	-	-	-	+	+	+	+	o	o	-	o	-	+

[a] Growth was scored as follows; **+** = strong growth   + = weak growth   - = no growth
o = death due to toxicity of compound

somewhat suprising because, at least in *Pseudomonas putida*, 15 genes are required to metabolize benzoate whereas only an additional 5 are required to convert toluene to benzoate (*1*). Two intermediates in toluene metabolism, benzyl alcohol and benzaldehyde, were either metabolized poorly or were toxic to most of the tested strains. Notable exceptions were five of seven strains of *Rhodococcus* that grew well on benzyl alcohol.

Cyclohexane did not support growth of any of the strains. This was not suprising as there are few well-documented reports of cyclohexane-utilizing microorganisms (*16, 17*). However, many of the strains were capable of utilizing cyclohexanone, a predicted early intermediate in cyclohexane metabolism (*18*). Other proposed intermediates, such as ε-caprolactone and adipate, supported good growth by most of the strains (data not shown).

Overall, the most versatile strains belonged to the genera *Mycobacterium*, *Nocardia* and particularly *Rhodococcus* - the so-called 'nocardioform' group of Actinomycetes. Although many strains of *Arthrobacter* were capable of growth on alkanes, few were capable of utilising other types of hydrocarbons. Within the nocardioform group, the Rhodococci appear to have the greatest biodegradation abilities. In addition to the strains characterized here, many other strains of *Rhodococcus* have been shown to degrade environmentally problematic compounds (*5 - 12*). It should also be noted that the taxonomy of this group of bacteria is still under review and many biodegrading strains previously identified as *Arthrobacter*, *Mycobacterium* or *Nocardia* have recently been reclassified as *Rhodococcus* (*19*).

Biochemical Analysis of Alkane Metabolism by *Rhodococcus erythropolis*. We chose *Rhodococcus erythropolis* ATCC 4277 as a model system to study alkane degradation by 'nocardioform' Actinomycetes because it is currently the only species of *Rhodococcus* that has been subjected to extensive genetic analysis (*20*). As shown in Table I, this strain is capable of utilizing *n*-alkanes ranging from undecane to heptadecane as sole sources of carbon and energy. It can also grow on benzoate, but is incapable of growth on any of the tested aromatic hydrocarbons, including toluene, xylene, napthalene and benzene. During subsequent analysis, we found that the strain could grow weakly on hexane when supplied at 0.1% v/v concentration rather than 0.4%.

To examine the enzymes involved in alkane metabolism by *R. erythropolis*, the strain was grown in RMM supplemented with 0.4% tetradecane as a sole carbon source, and a cell-free extract prepared as described in materials and methods. The extract was then assayed for a number of enzymes possibly involved in alkane metabolism using modifications of previously published protocols (*21*). A summary of the results of these assays and the metabolic pathway that could possibly be used by *R. erythropolis* is presented in Figure I. Using 1-octanol and octanal as substrates, we were able to find NAD^+ -dependent long-chain primary alcohol and aldehyde dehydrogenase enzyme activities similar to those described in other alkane-utilizing bacteria (*21, 22*). Both of these enzymes showed optimal activity against substrates with chain lengths from 6 to 10 carbon atoms (data not shown). However, we were also able to detect an extremely active NAD^+-dependent long-chain secondary alcohol dehydrogenase in the cell free extract. We have subsequently purified this enzyme and shown that it has optimal activity with 2-nonanol and is one of the most active bacterial long-chain secondary alcohol dehydrogenase yet described (Ludwig, B., A. Akundi and K. Kendall, manuscript submitted for publication.). The presence of this enzyme activity suggested that *R. erythropolis* may degrade alkanes via the subterminal oxidation pathway that has been described in certain fungi (*23*) and bacteria (*24 - 27*) as shown in Figure I. To investigate this possibility furthur, we used an indirect assay (described in materials and methods) to test for the presence of the predicted alkyl acetate esterase. Using this assay, we were able to demonstrate a slow octyl acetate-dependent reduction of NAD$^+$ in cell-free extracts. This appears to be due to an esterase converting octyl actetate to 1-octanol which is subsequently oxidized by the primary alcohol dehydrogenase. Unfortunately, we were unable to detect either of the predicted oxygenase enzymes in the pathway using a variety of different protocols.

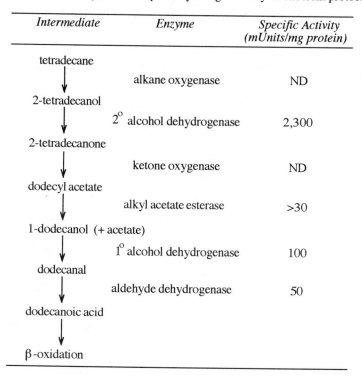

Intermediate	Enzyme	Specific Activity (mUnits/mg protein)
tetradecane		
↓	alkane oxygenase	ND
2-tetradecanol		
↓	2° alcohol dehydrogenase	2,300
2-tetradecanone		
↓	ketone oxygenase	ND
dodecyl acetate		
↓	alkyl acetate esterase	>30
1-dodecanol (+ acetate)		
↓	1° alcohol dehydrogenase	100
dodecanal		
↓	aldehyde dehydrogenase	50
dodecanoic acid		
↓		
β-oxidation		

Figure I. The Subterminal Oxidation Pathway for Alkane Metabolism.
The specific activities of enzymes found in cell extracts of *Rhodococcus erythropolis* ATCC 4277 are indicated in mUnits per mg protein. ND = not detected.

Isolation of Mutants of *Rhodococcus erythropolis* Defective in Alkane Metabolism. In order to further elucidate the pathway of alkane metabolism in *Rhodococcus*, cells of *R. erythropolis* were subjected to UV mutagenesis in an attempt to isolate mutants that could no longer utilize alkanes as sole sources of carbon and energy. From two independent experiments, a total of 6,000 mutagenized colonies were screened, resulting in the isolation of 21 mutants incapable of growth on either tridecane and/or tetradecane, but still capable of normal growth on sucrose. An additional 32 mutants were found that showed reduced growth on alkanes; these were not studied further. The 21 mutants were screened for their ability to grow on alkanes ranging in size from undecane to heptadecane. The results of this screening are summarized in Table II.

The mutants could be grouped into four general classes; Class I mutants failed to grow on any of the tested alkanes. Class II mutants were capable of growth on longer chain alkanes (generally tetradecane and above), but were incapable of growth on tridecane, dodecane or undecane. Conversely, Class III mutants could grow on short alkanes (undecane, dodecane and tridecane) but were incapable of

Table II. Alkane Utilization Profiles of Mutants of *Rhodococcus erythropolis* Defective in Alkane Degradation

Mutant	C6	C10	C11	C12	C13	C14	C15	C16	C17
				n-alkane[a]					
Wild Type	S	w	S	S	S	S	S	S	S
Class I									
A5	-	-	-	-	-	-	-	-	-
A8	-	-	-	-	-	-	-	-	-
B12	-	-	-	-	-	-	-	-	-
D7	-	-	-	-	-	-	-	-	-
H9	-	-	-	-	-	-	-	-	-
H13	-	-	-	-	-	-	-	-	-
I12	-	-	-	-	-	-	-	-	-
Class II									
A15	w	-	-	-	-	w	S	S	S
E2	w	-	-	-	w	S	S	S	S
F1	-	-	-	-	w	w	w	w	w
G19	S	-	-	-	-	-	w	S	w
I4	S	-	-	-	-	S	S	S	S
I6	S	-	-	-	w	S	S	S	S
I13	S	-	-	-	w	S	S	S	S
Class III									
B19	S	w	S	S	S	w	w	w	w
C7	w	-	S	S	S	w	-	-	-
C10	w	w	S	S	S	w	-	-	-
C20	w	-	S	S	S	w	w	w	w
Class IV									
D3	-	-	w	-	S	-	S	-	w
E7	w	-	-	-	S	w	S	w	S
G9	w	-	w	-	S	w	S	-	S

[a] S = strong growth, w = weak growth, - = no growth

growth on the longer chain alkanes. Class IV mutants were unusual in that they could utilize odd chain length alkanes (i.e. undecane, tridecane, pentadecane and heptadecane), but were incapable of growth on dodecane, tetradecane and hexadecane.

Growth of *R. erythropolis* on Predicted Alkane Metabolic Intermediates. The wild-type strain of *R. erythropolis* was screened for its ability to grow on various predicted intermediates in alkane metabolism. A summary of the results is presented in Table III. Many of the substrates appeared toxic to the strain at high concentration (0.4%) even though they served as growth substrates at lower (< 0.1%) concentrations. This effect was particularly noticeable with fatty acids and ketones. The strain was incapable of growth on any aldehydes, possibly due to severe toxicity. In general, short and medium chain-length compounds provided the best growth. Significant growth was found only with fatty acids and alcohols up to eight carbon atoms in length. Longer chain lengths were incapable of supporting growth and often appeared to be toxic to the strain (data not shown). With the exception of aldehydes, the strain was found to be capable of strong growth on at least some chain-lengths of all the predicted intermediates in the sub-terminal oxidation pathway.

TABLE III. Growth of *R. erythropolis* on Predicted Alkane Metabolism Intermediates

Compound Class	Number of carbon atoms[a]								
	C3	C4	C5	C6	C7	C8	C9	C10	C11
2-alcohols	w	w	w	S	S	S	-	-	-
2-ketones		-		S	S	S	S	S	S
alkyl acetates	S	S	w		S		S		
1-alcohols		S	S	S	w	w		-	-
fatty acids	S	S	S	S	S	w	w	w	

[a] S = strong growth, w = weak growth, - = no growth, (blank) = not tested.

Growth of Mutants on Alkane Metabolic Intermediates. Each of the mutants was tested for growth on predicted alkane pathway intermediates in an attempt to determine if the metabolic defects could be inferred from the substrate utilization profiles. Each mutant was screened for growth on the entire range of primary and secondary alcohols, alkyl acetates, 2-ketones and fatty acids that the wild type strain was capable of utilizing. A summary of these results for each of the mutant classes is presented below.

Class I (7 mutants). None of the Class I mutants were capable of growth on any of the alcohols, ketones or alkyl acetates tested. The mutants could be separated into three groups based on their fatty acid utilization profiles (Table IV). Two of the mutants (H9 and D7) were incapable of growth on fatty acids; these could possibly be defective in the β-oxidation pathway for fatty acid metabolism (*28*). One mutant (B12) grew well on all the fatty acids. It is possible that this mutant may be defective in either an alcohol or aldehyde dehydrogenase as it was also incapable of growth on alcohols (data not shown). However, we were able to detect apparently normal levels of both enzymes in the strain (data not shown). The remaining four mutants (A5, A8, H13 and I12) grew only on odd chain-length fatty acids. Such mutants could be predicted to be due to mutations affecting the glyoxylate bypass of the tricarboxylic acid cycle (*29*). We have subsequently shown that three of the four mutants are in fact deficient in isocitrate lyase, the first enzyme of the glyoxylate shunt (Vuong, L., and K. Kendall, manuscript in preparation).

Table IV. Growth of Class I Mutants on Fatty Acids

Mutant	Fatty acid chain length[a]								
	C2	C3	C4	C5	C6	C7	C8	C9	C10
Wild Type	S	S	S	S	S	S	w	w	w
Class Ia									
A5	-	S	-	S	-	w	-	S	-
A8	-	S	-	S	-	w	-	S	-
H13	-	S	-	S	-	w	-	-	-
I12	-	S	-	S	-	S	-	S	-
Class Ib									
B12	S	S	S	S	w	w	w	w	S
Class Ic									
H9	-	-	-	-	-	-	-	-	-
D7	-	-	-	-	-	-	-	-	-

[a] S = strong growth, w = weak growth, - = no growth.

Class II (7 mutants). All of the Class II mutants were capable of growth on fatty acids but gave varying reponses on the predicted alkane metabolic intermediates. Table V summarises the growth patterns of the Class II and Class III mutants on the different classes of intermediates. Three Class II mutants (A15, E2 and F1) could not grow on ketones, primary alcohols or alkyl acetate esters but were capable of weak growth on 2-hexanol. The remaining four mutants (I4, I6, I13 and G19) gave varying responses; these ranged from I4, which was identical to the wild type in its growth patterns on all of the intermediates, to G19, which failed to grow on any of the intermediates. No enzyme deficiencies were detected in any of the strains.

Table V. Growth Patterns of Class II and Class III Mutants on Predicted Alkane Metabolism Intermediates

Mutant	Intermediate Compounds[a]				
	2-alcohols	2-ketones	alkyl acetates	1-alcohols	fatty acids
Class II mutants					
A15	(6)	-	-	-	+
E2	(6)	-	-	-	+
F1	(6/7)	-	-	-	+
G19	-	-	-	-	(+)
I4	+	+	+	+	+
I6	+	-	-	(+)	+
I13	-	-	-	-	+
Class III mutants					
B19	(6)	+	-	+	+
C7	-	-	-	-	+
C10	-	(8)	-	(+)	+
C20	-	-	-	-	+

[a] Each mutant was tested for growth on the full range of chain-lengths of each type of compound that supported growth of the wild type strain. Growth was scored as follows; + = strong growth with the same utilization pattern as the wild type, (+) = weak growth with a wild type pattern of utilization, (6), (6/7), (8) = growth only on the chain length(s) indicated, - = no significant growth on any chain-length.

Class III (4 mutants).
The Class III mutants gave similar results to Class II in that all were capable of growth on fatty acids and gave varying results with the intermediates (Table V). C7 and C20 could not grow on any intermediates whereas B19 and C10 were capable of limited growth on primary alcohols and ketones but could not grow on secondary alcohols or alkyl acetate esters. Again, no enzyme defects could be found in any of the mutants.

Class IV (3 mutants).
Suprisingly, the Class IV mutants, which could only grow on odd chain-length alkanes, were all capable of substantial growth on both odd and even chain-length fatty acids. All were incapable of significant growth on any alcohols, ketones or alkyl acetate esters (data not shown). The only difference between the 3 mutants was that G9 and E7 were capable of growth on hexane, whereas D3 was not (see Table II). Levels of the four detectable alkane metabolic enzymes were unaffected in all three mutants. However, in agreement with the prediction that these mutants could also be affected in the glyoxylate bypass, two of the strains (G9 and E7) were

subsequently found to be defective in the regulation of isocitrate lyase (Vuong, L., and K. Kendall, manuscript in preparation).

Discussion

After screening a large number of Actinomycetes for their ability to utilize a wide range of hydrocarbons as sole sources of carbon and energy, we selected *Rhodococcus erythropolis* (ATCC 4277) as a model strain with which to investigate alkane biodegradation by the nocardioform group of Actinomycetes. This strain has previously been subjected to genetic analysis (20) and grows well on alkanes ranging in size from undecane to heptadecane, but grows only weakly on hexane and decane and is incapable of growth on heptane, octane and nonane. Although the strain will grow on tetradecane in the presence of 0.4% octane (K. Kendall, unpublished data.), suggesting that toxicity is not the cause of his effect, it has been noted that toxic interactions between medium chain-length alkanes and sensitive cells can be masked by the presence of long-chain, metabolizable alkane growth substrates (30).

When *R. erythropolis* was grown in minimal medium with tetradecane as the sole source of carbon and energy, we were able to detect four soluble enzyme activities related to alkane metabolism. NAD+-dependent long-chain primary alcohol and aldehyde dehydrogenases were expected to be present based on work in other bacteria (22). Most suprising was the discovery of a NAD+-dependent secondary alcohol dehydrogenase with high specific activity against 2-octanol. By indirect means, an alkyl acetate esterase activity was also detected in cell extracts. Taken together, these results suggest that the strain may metabolize alkanes using an initial subterminal oxidative attack such as that first described in fungi (23). Interestingly, although this metabolic pathway has not been well-studied, many of the bacteria thought to use it belong to genera closely related to *Rhodococcus* such as *Arthrobacter* (26, 27) and *Mycobacterium* (25).

Mutants of *R. erythropolis* defective in alkane utilization were readily obtained after UV mutagenesis. Of 21 'tight' mutants studied, only 7 were completely incapable of growth on alkanes. Suprisingly, four of these could grow on odd chain-length fatty acids, but were incapable of growth on even chain-length fatty acids. This raised the possibility that the mutants may be defective in the glyoxylate bypass, an anapleurotic pathway required for growth on acetate and other two carbon compounds, including the products of fatty acid metabolism (reviewed in 28). We have subsequently demonstrated that three of these four mutants are deficient in isocitrate lyase (Vuong, L., and K. Kendall. manuscript in preparation). Of the remaining three Class I mutants, two were incapable of growth on fatty acids, suggesting a defect in fatty acid metabolism, and one was capable of growth on all fatty acids. It is possible that this mutant (B12) is truly defective in an important step in alkane metabolism. However, no defects in alcohol or aldehyde dehydrogenases were found in this strain.

The three mutants belonging to Class IV were defective in utilization of even chain-length alkanes only. Although two of these mutants have subsequently been shown to be defective in isocitrate lyase (Vuong, L., and K. Kendall, manuscript in preparation), all three mutants were capable of substantial growth on both even and odd chain-length fatty acids. The reasons behind their ability to grow on even chain-length fatty acids are unclear.

The remaining 11 mutants gave extremely variable growth profiles. They could be separated into 'short-chain' mutants (Class II) and 'long-chain' mutants (Class III) based on their inability to grow efficiently using *n*-alkanes of less than C_{14} or greater than C_{13} respectively. These mutants may be similar to some of the mutants of '*Mycobacterium rhodochrous*' described by Jenkins et al. (5). They also obtained large numbers of mutants that were not completely incapable of growth on alkanes, but rather had modified patterns of utilization. In *Pseudomonas putida*, it has been shown that such mutants can be due to strains containing multiple, redundant, alcohol and aldehyde dehydrogenases, each with different chain-length specificities (*31*). The mutants we obtained also gave extremely variable results when screened for their growth patterns on predicted intermediates of alkane metabolism. No clear pattern or obvious enzymatic defects emerged from these growth studies.

The mutant analysis presented in this study suggests that alkane metabolism in *R. erythropolis* may be different from many other bacteria. Similar studies in *Pseudomonas* (*31-33*) yielded relatively simple mutant classes with clear defects in specific steps in metabolism. Seven of the twenty one mutants we isolated appeared to have defects in the glyoxylate bypass pathway. The fact that these mutations were found at such high frequency, plus the large number of mutants found with modified patterns of hydrocarbon utilization, suggests that there may be a number of pathways and/or redundant enzymes used by *R. erythropolis* for growth on alkanes.

Acknowledgements

This work was supported by DoD grant number 2-89/116/88-150 awarded to Tulane/Xavier Universities and administered through the Tulane and Xavier Center for Bioenvironmental Research.

Literature Cited

1. Harayama, S.; Timmis, K.N. In *Genetics of Bacterial Diversity*; Hopwood, D.A.; Chater, K.F., Eds.; Academic Press, Inc. (London), Ltd: London, United Kingdom. 1989; pp. 151-174.
2. Kok, M.; Oldenhuis, R.; van der Linden, M. P.; Meulenberg, C. H.; Kingma, J.; Witholt, B. *J. Biol. Chem.* **1989**, *264*:5442-51.
3. Hayase, N.; Taira, K.; Furukawa, K. *J. Bacteriol.* **1990**, *172*:1160-1164.
4. Kurkela, S.; Lehvaslaiho, H.; Palva, E. T.; Teeri, T. H. *Gene* **1988**, *73*:355-62
5. Jenkins, P.G.; Raboin, D.; Moran, F.. *J. Gen. Microbiol.* **1972**, 72:395-398.
6. Grund, E.; Denecke, B.; Eichenlaub, R.. *Appl. Environ. Microbiol.* **1992**, *58*:1874-1877.
7. Bernhardt, D.; H. Diekmann. *Appl. Microbiol. Biotechnol.* **1991**, *36*:120-123.
8. Lee, S.T.; Lee, S-P.; Park, Y-H. *Appl. Microbiol. Biotechnol.* **1991**, *35*:824-829.
9. Ikehata, O.; Nishiyama, M.; Horinouchi, S.; Beppu, T. *Eur. J. Biochem.* **1989**, *181*:563-570.
10. Fuchs, K.; Schreiner, A.; Lingens, F.. *J. Gen. Microbiol.* **1991**, *137*:2033-2039.
11. Haeggblom, M. M.; L. J. Nohnek; M. S. Salkinoja-Salonen. *Appl. Environ. Microbiol.* **1988**, *54*:3043-3052.
12. Dabrock, B.; Riedel, J.; Bertram, J.; Gottschalk, G. *Arch. Microbiol.* **1992**, *158*:9-13.
13. Hopwood, D. A.; Bibb, M. J.; Chater, K. F.; Kieser T.; Bruton, C. J.; Kieser, H., Lydiate, M. D. J.; Smith, C. P.; Ward, J. M.; Schrempf, H. *Genetic manipulation of Streptomyces - a laboratory manual*. The John Innes Foundation: Norwich, United Kingdom. 1985.
14. Bradford, M. M. *Anal. Biochem.* **1976**, *72*:248-254.
15. Hopwood, D.A.; Kieser T.; Wright, H.M.; Bibb, M.J. *J. Gen. Microbiol.* **1983**, *129*:2257-2269.
16. Stirling, L.A.; Watkinson, R.J.; Higgins, I. J. *J. Gen. Microbiol.* **1977**, 99:119-125.
17. Anderson, M.S.;Hall, R.A.; Griffin, M. *J. Gen. Microbiol.* **1980**, *120*:89-94.

18. Perry, J.J. In *Petroleum Microbiology*; Atlas, R.M., Ed; Macmillan Publishing Co.: New York, NY,1987; pp. 61-97.
19. Tsukamura, M.; Mizuno, S.; Tsukamura, S.; Tsukamura, J. *Int. J. Syst. Bacteriol.* **1979**, *29*:110-129.
20. Brownell G. H.; Denniston, K. In *The Biology of Actinomycetes*; Goodfellow, M.; Mordarski, M.; Williams, S.T., Eds.; Academic Press, Inc. (London), Ltd: London, United Kingdom, 1984. pp. 201-208,
21. Fox, M.G.A.; Dickinson, F.M.; Ratledge, C. *J. Gen. Microbiol.* **1992**, *138*:1963-1972.
22. Singer, M. E.; Finnerty, W. R. In *Petroleum Microbiology*; Atlas, R.M. Ed.; Macmillan Publishing Co.: New York, NY, 1987; pp. 1-59.
23. Allen, J. E.; Markovetz, A. J.. *J. Bacteriol.* **1970**, *103*:426-434.
24. Forney, F. W.; Markovetz, A. J.. *J. Bacteriol.* **1968**, *96*:1055-1064.
25. Fredricks, K. M. *Antonie van Leeuwenhoek J. Microbiol. Serol.* **1967**, *33*:41-48.
26. Klein, D. A.; Davies, J. A.; Casida, L. E. Jr. *Antonie van Leeuwenhoek J. Microbiol. Serol.* **1968**, *34*:495-503.
27. Klein, D. A.; Henning, F. A. Appl. Microbiol. **1969**, *17*:676-681.
28. Nunn, W. D. 1987. In *Escherichia coli and Salmonella typhimurium*; Ingraham, J. L.; Low, K. B.; Magasanik, B.; Schaechter, M.; Umbarger, H. E. Eds.; American Society for Microbiology: Washington, D.C, 1987; pp. 285-301.
29. Chung, T.; Klumpp, D. J.; LaPorte, D. C. *J. Bacteriol.* **1988**, *170*:386-392.
30. Gill, C.O. and C. Ratledge. *J. Gen. Microbiol.* **1972**, *72*:165-172.
31. Grund, A.; Shapiro, J.; Fennewald, M.; Bacha, P.; Leahy, J.; Markbreiter, K.; Nieder, M.; Toepfer, M. *J. Bacteriol.* **1975**, *123*:546-556.
32. Macham, L. P.; Heydeman, M. T. *J. Gen. Microbiol.* **1974**, *85*:77-84.
33. Illarionov, E. F. *Mikrobiologiya* **1972**, *41*:126-131.

RECEIVED March 14, 1995

Chapter 21

Determination of Bioavailability and Biodegradation Kinetics of Polycyclic Aromatic Hydrocarbons in Soil

Henry H. Tabak[1], Rakesh Govind[2], Chao Gao[2], Lei Lai[2], Xuesheng Yan[2], and Steven Pfanstiel[2]

[1]National Risk Management Research Laboratory, U.S. Environmental Protection Agency, 26 West Martin Luther King Drive, Cincinnati, OH 45268
[2]Department of Chemical Engineering, University of Cincinnati, Cincinnati, OH 45221

Biological treatment systems, especially when used in conjunction with physical /chemical treatment methods, hold considerable promise for efficient, safe, economical on-site and *in-situ* treatment of polycyclic aromatic hydrocarbons (PAHs). Fundamental understanding of biodegradation kinetics and factors that control the rate of biodegradation can provide insight into the "optimal" range of environmental parameters for improvement of microbiological activity and enhancement of contaminant biodegradation rates in soil systems. This paper reports on a study of adsorption /desorption kinetics and equilibria of four polycyclic aromatic hydrocarbons in soil slurry systems. Respirometric studies were conducted to quantify biodegradation kinetics in soil slurry reactors. Abiotic adsorption and desorption rates, cumulative oxygen consumption and carbon dioxide evolution data were used with non-linear regression techniques to determine the biokinetic parameters using a detailed mathematical model.

Polycyclic aromatic hydrocarbons (PAHs) are toxic and hazardous chemicals regulated by the U.S. Environmental Protection Agency as priority pollutants. There is interest in understanding the fate of these compounds in soil-water systems and determining the mechanism and rate of biodegradation. Various studies have identified specific organisms capable of degrading PAH compounds (1,2,3,4). Information on aerobic pathways is generally limited to two-and three-ring PAH compounds including naphthalene (5,6), acenaphthene (4), and phenanthrene (5,7). Literature on microbial degradation and stability of PAH compounds in soil-water systems has been summarized by Atlas (8) and Sims and Overcash (9). Most previous studies did not specifically consider abiotic and biotic mechanisms for decrease of aqueous-phase PAH concentrations in soil-water systems. PAH contamination is found in petroleum and wood-preserving wastes and land treatment techniques have been applied for treating these wastes (10,11,12). Studies have documented 70% decreases in concentration of carcinogenic PAHs in less than 100 days in some land treating units (13).

In this paper, systematic studies have been conducted on the adsorption and desorption of four PAH compounds: naphthalene, acenaphthylene, acenaphthene, and phenanthrene. In these studies both the kinetics and equilibria compositions were determined at various initial concentrations. Studies were also conducted to determine the biodegradation kinetics in soil slurry reactors using respirometric

0097–6156/95/0607–0264$12.00/0

techniques. The values of the Monod kinetic parameters were determined from the oxygen uptake data using non-linear regression methods.

Background

The adsorption and desorption of hydrophobic PAH compounds is rate limited by intrasorbent diffusion. Diffusion-limited sorption has been proposed by many researchers as a probable cause of non-equilibrium for organic solutes (14, 15). Wu and Gschwend (16, 17) studied sorption kinetics of hydrophobic organic compounds by natural sediments and soils. A retardation factor was proposed to reflect microscale partitioning of the compound between the intra-aggregate pore fluids and the soils constituting the aggregate grains.

Studies by Voice and Weber (18) quantified adsorption of hydrophobic pollutants by sediments, soils and suspended solids. In recent studies (19), contaminant sorption by soil and sediments was described as a multiple reaction phenomenon, and a composite reactivity model was introduced to characterize intrinsic heterogeneities in soil systems.

Several models have been used to describe the equilibrium distribution (isotherm) of contaminant between the solid and solution phases. The simplest model is a linear isotherm (20, 21) which is usually valid at contaminant concentrations below 0.056M. At higher contaminant concentrations, either the Langmuir or Freundlich isotherm models have been used in the literature. Studies have also been conducted to correlate the isotherm parameters with soil characteristics, such as soil organic carbon content. Studies by Means (22) on sorption properties of nitrogen heterocyclic and substituted PAH compounds indicate that sorption can be determined by the organic carbon content of the soil/sediment, as originally suggested by Karickhoff (23), and is independant of substrate pH, cation exchange capacity, textured composition or clay mineralogy.

Non-equilibrium sorption of hydrophobic organic compounds by natural sorbents were reported by Brusseau et al. (24) and Brusseau and Rao (25) and compared with rate data obtained from systems wherein rate-limited sorption was caused by specific sorbate-sorbent interactions. The comparison showed that intra-organic matter diffusion is responsible for non-equilibrium sorption.

Adsorption and desorption of organic hydrophobic compounds, such as PAHs, not only impacts the compound's transport in soil and soil-water ecosystems, but also their bioavailability and ultimate persistence in the environment. Studies by Weissenfels et al. (26) on the influence of adsorption of PAHs by soil particles have shown that migration of PAHs into soil organic matter representing less accessible sites within the soil matrix decreases the compound's bioavailability (biotoxicity) and lowers the extent of overall biodegradation of the PAH compounds. Furthermore, the extent of bioavailability of sorbed PAHs depends on the organism, which makes it difficult to generalize bioavailability of sorbed substrates.

Microbial degradation of PAHs under aerobic, anaerobic, and denitrification conditions in soil-water systems have also been examined (27) and compared with chemical degradation in the presence of manganese oxide. Results of this study showed that low-molecular weight unsubstituted PAHs are amenable to microbial degradation in soil-water systems under denitrification conditions. Respirometric studies on microbial degradation of PAHs in contaminated soil suspensions were performed using specialized mixed culture Sapromat respirometer (12) to demonstrate soil decontamination.

Methodology

Establishment of Soil Microcosm Reactors.   A microcosm reactor is an airtight rectangular container (20 in. x 12 in. x 12 in.) constructed of glass supported by stainless steel panels in which

Table I. Characteristics of soil used in this study.

Soil moisture	17%
Organic Matter	2.9%
Classification	silt loam
Particle size distribution	
Clay (< 2 μm)	24%
Silt (2-44 μm)	58%
Sand (> 44μm)	18%
Cation exchange capacity	6.5
Soil pH	6.1
Bulk density	1.06
1/3 bar moisture	30.7%
15 bar moisture	14.8%
Nutrients in soil (ppm)	
Phosphorus	17(h)
Potassium	90(m)
Magnesium	80(l)
Calcium	1100 (m)
Sodium	17(v)

Note: h = high, m = medium, l = low, v = very low

undisturbed uncontaminated forest soil was placed. The soil characteristics, measured using standard methods, are summarized in Table I. The nutrients and appropriate contaminants are sprayed from the top using liquid atomizing sprays. The bottom of the reactor is equipped with ports to allow the drainage of leachates. A controlled flowrate of carbon dioxide-free air is passed through the reactor and the exit gas is bubbled through potassium hydroxide solution to quantify the average evolution rate of carbon dioxide in the reactor. Figure 1 shows a schematic of a microcosm reactor system.

Each microcosm reactor represents a controlled site, which eventually selects out the acclimated indigenous microbial population in the soil for the contaminating organics. Samples of soil are then taken from the microcosm reactors and used as source of acclimated microbial inoculum for measuring: (1) oxygen uptake respirometrically; (2) carbon dioxide generation kinetics in shaker flask reactors; and (3) for studies with other soil reactor systems. The microcosm reactor units are also being used directly to evaluate the biodegradability of the pollutant organics and to measure their average biodegradation rate in this intact, undisturbed soil bed.

A microcosm reactor was contaminated with 25 ppm each of several PAHs dissolved in a mixture of deionized-distilled water and a 0.5% solution of a surfactant, Triton X-100. Control microcosm reactors consisted of uncontaminated soil sprayed with an equal amount of deionized-distilled water and a second reactor sprayed with 0.5% solution of surfactant, Triton X-100.

Measurement of Adsorption Equilibria and Kinetics. Soil adsorption kinetics and equilibria are measured using batch well-stirred bottles. The soil is initially air dried and then sieved to pass a 2.00 mm sieve. 50 g of soil sample is placed in each bottle and mixed with 250 mL of distilled deionized water containing various concentrations of the compound and 1 ml of mercuric chloride saturated solution to minimize biodegradation. The soil:solution ratio is expressed as the oven-dry equivalent mass of adsorbent in grams per volume of solution.

The liquid is sampled after 2, 4, 6, 8, 10, 12, 14, 16, 18, 20, and 36 hours. After the predefined time has elapsed, the bottle contents are centrifuged and the liquid sample is taken using a syringe connected to a 0.45 μm porous silver membrane filter. The filter prevents soil particles from entering the sample.

The concentration of the chemical compound in the liquid sample was analyzed using three methods: (1) standard extraction (EPA methods 604 and 610) with methylene chloride followed by GC/MS analysis; (2) HPLC analysis; and (3) scintillation counting of the ^{14}C using radiolabelled compound. All three analytical methods were calibrated using standard stock solutions of each compound. The calibration data are used to convert the peak areas (GC/MS or HPLC) or counts of disintegrations per minute (DPM) for radiolabelled compounds to the actual liquid concentration.

To obtain the calibration curve, each PAH chemical was weighed equal to 80% of standard solubility in hexane. The weighed chemical was dissolved completely in hexane. Standard solutions containing 60%, 40%, 20% and 10% standard solubility levels were obtained by diluting with hexane. A HPLC column (Supelco LC-PAH 15 cm x 4.6 cm) was used to obtain peak areas for each standard solution. A plot of peak areas versus the concentration was made to obtain the standard calibration curve for each chemical. Verification of analytical accuracy with an internal standard showed that the above procedure was accurate to the same degree for the compounds being studied. For chemicals with extremely low solubility (< 1 ppm), radiolabelled compounds were used with scintillation counting. Standard solutions were made in the same manner and calibrated with counts of disintegrations per minute (DPM).

Stock solutions of each PAH chemical were made by first weighing chemical equal to at least twice standard solubility level in water. The chemical was thoroughly mixed with ultrapure water using teflon coated magnetic stir bar for 48 hours. The solution was allowed to settle for 48 hours. The

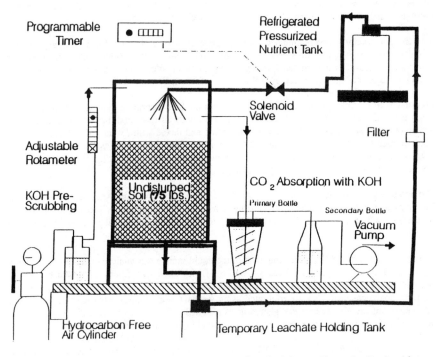

Figure 1. Schematic of a microcosm reactor system for obtaining acclimated soil microbiota.

saturated solution was decanted and filtered using a 5 μm millipore filter to remove all suspended particles. The liquid concentration was analyzed using either HPLC or scintillation counting. Other stock solutions were obtained by diluting with ultrapure water to obtain solutions with 80%, 60%, 40%, 20% and 10% standard solubility levels.

HPLC had the distinct advantage of requiring no extraction with a solvent as in the case of GC/MS analysis. However, HPLC method did not have enough sensitivity for liquid phase concentrations below 1 ppm. Solvent extraction was time consuming, used excessive amounts of solvent and often did not yield 100% recovery of the compound. ^{14}C scintillation counting required radiolabelled compound but was fast and easy to use. However, limited number of ^{14}C compounds are available and radiolabelled compounds are expensive. Furthermore, scintillation counting methods result in the generation of radioactive waste that requires proper handling and disposal.

For liquid phase concentrations below 1 ppm, ^{14}C scintillation counting is the preferred method. For higher concentrations, and for compounds with low octanol-water partition coefficient, HPLC is better than extraction with organic solvent and GC/MS analysis. In this study, results from all three analytical methods agreed closely.

From the initial amount of compound and analysis of the liquid phase, the amount of compound adsorbed in the soil is obtained by difference. Equilibrium is defined when the liquid concentration reaches a stationary value, which is usually attained in 24 hours.

The adsorption kinetics was analyzed using the following equation:

$$dS_w/dt = k_{1a} S_w^{m_a} - k_{-1a} S_s \tag{1}$$

where S_w is the compound concentration in the liquid phase, S_s is the compound concentration in the soil phase and is equal to (X/M) at any time, t, and k_{1a}, k_{-1a} and m_a are the kinetic parameters for adsorption.

The adsorption isotherm for each compound was analyzed using the Freundlich isotherm equation:

$$X/M = k_a C^{1/na} \tag{2}$$

where X is the amount of chemical adsorbed in soil (mg), M is the mass of soil (Kg), k_a is the equilibrium constant indicative of adsorptive capacity $(mg/Kg)(L/mg)^n$, C is the equilibrium solution concentration after adsorption (mg/L) and n is a constant indicative of adsorption intensity.

Measurement of Desorption Kinetics and Equilibria. Desorption studies were conducted by first adsorbing the chemical in the soil until equilibrium is achieved. Then desorption studies are conducted using two methods: (1) centrifuging the soil slurry to separate the liquid from the soil phase, withdrawing a specified volume of the liquid and adding deionized distilled water; and (2) diluting with deionized distilled water. In both methods, the concentration in the liquid phase is measured to obtain the desorption kinetics and equilibria.

In the first method, 100 ml of deionized distilled water is mixed with 20 grams of soil and specified concentration of chemical for adsorption. After adsorption equilibrium is attained, the whole solution is centrifuged and 90 ml of supernatant is taken out and replaced with an equal volume of deionized distilled water and 1 ml of mercuric chloride saturated solution to inhibit biodegradation. Desorption begins from this time. 20 ml sample is withdrawn at 4, 8, 16, 24, 48, 72, 96 and 120 hours. Each sample is withdrawn from a separate bottle. Each sample is filtered with 0.45 μm filter and extracted with methylene chloride and analyzed using GC/MS technique to determine the concentration of solute. HPLC analysis and ^{14}C scintillation counting analysis for radiolabelled compounds were also used as additional methods for comparative purposes.

In the second method, 250 ml of deionized distilled water is mixed with 50 gms of soil and specified concentration of chemical for adsorption. After adsorption equilibrium is attained, the sample is diluted with an equal volume of deionized distilled water and with 1 ml of mercuric chloride saturated solution to inhibit biodegradation. 20 ml sample is withdrawn at 4, 8, 16, 24, 48, 72, 96 and 120 hours. Each sample is withdrawn from a separate adsorption bottle. The sample was analyzed using extraction with methylene chloride followed by GC/MS analysis. HPLC analysis and ^{14}C scintillation counting analysis for radiolabelled compounds were also used as additional methods to compare the analysis results in desorption studies. Results from all three analytical approaches agreed closely, and subsequently HPLC was used exclusively.

It was found that significant errors resulted when the first method using centrifugation of the soil slurry was used. This was mainly due to difficulties in separating colloidal size particles using the centrifugation, and significant amount of the chemical was adsorbed on these micron size particles. Attempts in using smaller pore size filters failed since significant pressure was necessary to force the liquid through the filter. Hence, the dilution method, which required no separation, was selected in our studies.

The desorption kinetics was analyzed using an equation similar to that for adsorption kinetics:

$$dS_w/dt = k_{1d} S_w^{m_d} - k_{-1d} S_s \qquad (3)$$

where k_{1d}, k_{-1d} and m_d are the kinetic parameters for desorption.

The desorption isotherm for each compound was also analyzed using the Freundlich equation:

$$X/M = k_d C^{1/n_d} \qquad (4)$$

where X is the equilibrium amount of chemical adsorbed in soil (mg), M is the mass of soil (Kg), k_d is the equilibrium constant indicative of desorption capacity $(mg/Kg)(L/mg)^n$, C is the equilibrium solution concentration after desorption (mg/L) and n is a constant indicative of desorption intensity.

Measurement of Bacterial-Soil Sorption Equilibria. The adsorption isotherm for the bacterial cells is determined by incubating soil microbiota with radiolabelled napthalene in a respirometric reactor until an oxygen uptake plateau is obtained, indicating that all napthalene has biodegraded either into $^{14}CO_2$, which is adsorbed in the KOH solution and into ^{14}C biomass. The soil suspension is allowed to settle for about 30 minutes. One ml of supernatant is sampled and the ^{14}C activity is measured by liquid scintillation counting. Equilibrium amounts of the ^{14}C biomass adsorbed to the soil is determined by subtracting from the total ^{14}C added initially, the ^{14}C present in the biomass in suspension and the ^{14}C present as carbon dioxide absorbed in the KOH solution.

A linear isotherm was found to adequately fit the experimental data on bacterial-soil sorption equilibria. The isotherm can be expressed as follows:

$$X_s = K_b X_w \qquad (5)$$

where X_s is the microbiota concentration in soil at equilibrium, X_w is the equilibrium microbiota concentration in the water phase and K_b is the linear bacterial sorption isotherm equilibrium parameter.

Respirometric Studies with Soil Slurries. Studies were conducted with soil slurry reactors, wherein the oxygen uptake was monitored respirometrically. The extent of biodegradation and the Monod kinetic parameters for variety of the organic pollutant compounds by soil microbiota were determined from oxygen uptake data. Various concentrations of soil (2, 5, 10%) and compound (50, 100 mg/L) were mixed with a synthetic medium consisting of inorganic salts, trace elements and either a vitamin solution or solution of yeast extract and stirred in the respirometric reactor flasks. Each PAH compound was weighed and added to the reactor flask as a finely divided powder. Due to low water solubilities of the PAH compounds, most of the compound initially remained as a precipitate at the bottom of the flask, but eventually emulsified as biodegradation proceeded in the flask. The flasks were connected to the oxygen generation flask and pressure indicator cells of a 12 unit Voith Sapromat B-12, electrolytic respirometer, and the oxygen uptake (consumption) data were generated as oxygen uptake velocity curves. A detailed description of the Voith Sapromat B-12 electrolytic respirometer (Voith Inc., Heidenheim, Germany) has been presented elsewhere (28).

Nutrients needed for microbial growth were added with the following initial concentrations present in each reactor flask (29): KH_2PO_4 (85 mg/L)), K_2HPO_4 (217.5 mg/L)), $Na_2HPO_4.2H_2O$ (334 mg/L), NH_4Cl (25 mg/L), $MgSO_4.7H_2O$ (22.5 mg/L), $CaCl_2$ (27.5 mg/L) and $FeCl_3.6H_2O$ (0.25 mg/L), $MnSO_4.H_2O$ (0.0399 mg/L), H_3BO_3 (0.0572 mg/L), $ZnSO_4.7H_2O$ (0.0428 mg/L), $(NH_4)_6Mo_7O_{24}$ (0.0347 mg/L), $FeCl_3.EDTA$ (0.1 mg/L), and yeast extract (0.15 mg/L). The soil served as a source of inoculum. The concentration of forest soil in the reactor flask varied from 2 to 10 % by weight, using dry weight of soil as the basis. The total volume of the slurry in the flask was 250 ml.

The test and control compound concentrations in the media were 100 mg/L. Aniline was used as the biodegradable reference compound at a concentration of 100 mg/l. The typical experimental system consisted of duplicate flasks for the reference substance, aniline, and the test compounds, a single flask for the physical/chemical test (compound control), and a single flask for toxicity control. The contents of the reaction vessels were preliminary stirred for an hour to ensure endogenous respiration state at the initiation of oxygen uptake measurements. Then the test compounds and aniline were added to it. The reaction vessels were then incubated at 25°C in the dark (enclosed in the temperature controlled waterbath) and stirred continuously throughout the run. The microbiota of the soil samples used as an inoculum were not pre-acclimated to the substrates. The incubation period of the experimental run was 28 to 50 days.

Kinetic Analysis of Soil Slurry Oxygen Uptake Data. The experimental data were analyzed using a mathematical model (equations given below) in conjunction with the effect of chemical adsorption/desorption on the soil particles (equations 1,2,3,4) and the adsorption / desorption of the microorganisms from the soil (equation 5):

$$O_2 = (S_{to} - S_t) - (X_t - X_{to}) - (S_p - S_{po}) \tag{6}$$

$$dX_t/dt = (\mu_w X_w S_w)/(K_w + S_w) + \mu_{sw} X_s S_s/(K_{sw} + S_s) - (bK_w X_t)/(K_w + S_t) \tag{7}$$

$$dS_t/dt = -(1/y)(\mu_w X_w S_w)/(K_w + S_w) - (1/y)(\mu_{sw} X_s S_s/(K_{sw} + S_s)(w/v) \tag{8}$$

$$dS_p/dt = -y_p(dS_t/dt) \tag{9}$$

where subscripts t, s, w and p represent the total, soil, water and degradation products respectively. O_2 is the cumulative oxygen uptake, S is the concentration of compound, X is the concentration of biomass, and S_p is the concentration of the degradation products. μ_w, K_w, y and y_p are the Monod equation maximum specific growth rate parameter, Michaelis constant, biomass yield and product yield coefficient, respectively for the water phase. μ_{sw}, K_{sw} are the Monod equation maximum specific growth rate parameter and Michaelis constant, respectively for the soil phase. b is the biomass decay coefficient in the water phase, t is time, w is the weight of soil in the slurry reactor and v is the total volume of water in the reactor. Experiments conducted earlier had determined that decay coefficient of biomass in the soil phase is negligible compared to decay rates in the water phase.

The experimental values of the oxygen uptake were matched with the theoretically calculated values and the best fit for the parameters were obtained using a standard adaptive random search method.

Measurement of Carbon Dioxide Evolution Rates. Carbon dioxide generation rates were measured in shaker-flask soil slurry systems and in the electrolytic respirometry soil-slurry reactors in order to assess the rate and extent of biodegradation/mineralization. The CO_2 generation rate measurement serves as an additional tool for quantitating the biofate of these organics in addition to the cumulative respirometric oxygen uptake data from which the biokinetics were derived. The CO_2 generation rate measurement experiments were performed using both the shaker flask and respirometric reactors to determine the compatibility and reproducibility between the data on CO_2 production in both systems. KOH solution was used to absorb the carbon dioxide produced in each reactor flask and the amount of carbon dioxide was quantitated by measuring the pH change of the KOH solution. A computer program was developed to solve the carbonate and bicarbonate equilibria equations for absorption of carbon dioxide in KOH solution and calculate the dependance of amount of carbon dioxide absorbed and changes in pH.

Measurement of carbon dioxide evolution with uniformly labelled ^{14}C PAHs were also conducted to determine the net carbon dioxide produced from the chemical and the carbon dioxide produced due to mineralization of soil organic matter.

Results and Discussion

Studies on Soil Microcosm Reactors. With the use of these specially designed microcosm reactors, it was possible to acclimate the indigenous microbiota to each class of compounds. Soil samples from the microcosm reactors are used as a source of acclimated microbiota for measuring oxygen uptake respirometrically to determine biodegradation kinetics and to determine carbon dioxide generation kinetics.

It should be emphasized that the soil microcosm reactors were used to obtain acclimated soil and measurements of average carbon dioxide evolution rates was used primarily to monitor the on-set of biodegradation of the contaminants by quantifying the net increase in carbon dioxide evolution over the control microcosms.

Analysis of Oxygen Uptake Data. Oxygen uptake data was generated with 5% soil slurry and the following polycyclic aromatic hydrocarbons (PAHs): acenaphthene, acenaphthylene, naphthalene and phenanthrene. 100 mg/L concentration of each PAH was used in the experiments. Initially the PAH compound was insoluble in the nutrient mixture and remained suspended as the mixture was mixed with the soil. However, after some acclimation time, the PAH compound was emulsified in the aqueous mixture and a separate insoluble precipitate could not be distinguished visually in the reactor flask.

Figure 2 shows the experimental oxygen uptake data and model calculated curve for 100 mg/L of naphthalene using 5% soil suspension. The acclimation time (not shown in the figure) during which no oxygen uptake occurred was 42 hours. The cumulative oxygen uptake data attained a plateau after 12 hours. Figure 3 shows the representative cumulative oxygen uptake data and model calculated curve for 150 mg/L of phenanthrene using 5% soil suspension. The acclimation time (not shown in the figure) at 150 mg/L of initial concentration was about 200 hours, and oxygen uptake attained a plateau after 40 hours.

The oxygen uptake data were analyzed using the mathematical model and the best-fit Monod parameters for the PAHs are given in Table II. The Monod kinetic parameter, μ_w, represents aqueous degradation of the dissolved or emulsified compound by the microorganisms present in the aqueous phase. The maximum specific growth rate parameter, μ_w, ranges from 0.127 to 0.482 for the four PAH compounds, which is typical for aerobic aqueous biodegradation of most organic compounds. The yield coefficient, y, is approximately 0.5 which is also typical for aerobic biodegradation. The Monod kinetic parameter, μ_{sw}, for the microorganisms immobilized in the soil matrix, is significantly higher than that for the aqueous phase. The approximate first-order kinetic parameter for the immobilized biomass (μ_{sw}/K_{sw}) is lowest for phenanthrene and highest for naphthalene. This agrees with the fact that phenanthrene with three fused aromatic rings is more difficult to biodegrade than naphthalene with two fused aromatic rings.

Analysis of the oxygen uptake data with adsorption/desorption kinetics showed that both adsorption and desorption attained equilibrium in less than 20 hours, while biodegradation usually involved acclimation times exceeding 20 hours. Hence, incorporating adsorption/desorption kinetics with cumulative oxygen uptake was equivalent to assuming adsorption/desorption equilibria during the biodegradation phase.

Measurement of CO_2 Generation Rate in Shaker Flask Slurry Systems. Carbon dioxide evolution results for PAHs are shown in Figure 4 for naphthalene and Figure 5 for phenanthrene. Acclimated soil or acclimated soil liquid, obtained by filtering a suspension of acclimated soil sample, were used in the experimental studies. Several concentrations of the compound ranging from 0 mg/L to 150 mg/L were used with a 10% soil slurry suspension. In general, the carbon dioxide evolution increased with compound concentration. Differences in carbon dioxide evolution amounts between the flasks with addition of acclimated soil and acclimated soil liquid is unclear at this time. Several possible explanations can be given which includes the fact that soil-liquid may have contained colloidal suspension of soil particles and/or microorganisms which would impact the extent and rates of biodegradation in the reactor systems.

The relationship of the CO_2 production in shaker flask soil slurry and respirometric soil slurry reactor systems was studied through measurement of CO_2 trapped by soda lime in respirometric vessels at various sampling times throughout the course of the respirometric oxygen uptake run. Measurements of cumulative CO_2 generation in the respirometric reactors were made using soda lime as absorbent initially and subsequently with the use of KOH solution which was periodically measured for the pH change.

Comparisons were made between the cumulative CO_2 concentrations experimentally measured for the different soil and compound concentrations in the shaker flask soil slurry systems and respirometric reactor soil slurry systems measuring oxygen uptake data. Ratio values were developed for the CO_2 experimentally measured in respirometer and shaker flask to the CO_2 calculated from oxygen uptake data.

The compatibility between the data on the CO_2 production in the shaker flask and respirometric vessel tests has been established and verified and the measurement of CO_2 evolution in shaker flask soil slurry systems was shown to be dependable for quantitative CO_2 analysis. The results on the

Table II. Biokinetic parameters for Polycyclic Aromatic Hydrocarbons (PAHs) from soil slurry experimental oxygen uptake data.

Kinetic Parameters	Acenaphthene	Acenaphthylene	Naphthalene	Phenanthrene
μ_w (1/hr)	0.127±0.002	0.372±0.003	0.229±0.001	0.482±0.002
K_w (mg/L)	0.813±0.02	5.90±0.03	2.478±0.03	18.63±0.03
μ_{sw} (1/hr)	2.818±0.002	4.659±0.003	15.04±0.001	4.895±0.002
K_{sw} (mg/L)	0.139±0.02	1.03±0.02	0.749±0.03	19.67±0.03
y (mg/mg)	0.459±0.001	0.425±0.001	0.423±0.001	0.386±0.001
y_p (mg/mg)	0.00069	0.00014	0.00068	0.00082
b (1/hr)	0.0047	0.0061	0.0036	0.0097

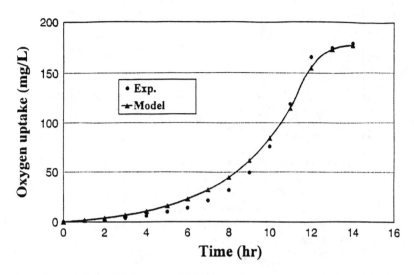

Figure 2.   Experimental data and model fit of cumulative oxygen uptake for 100 mg/L of naphthalene using 5% soil suspension.

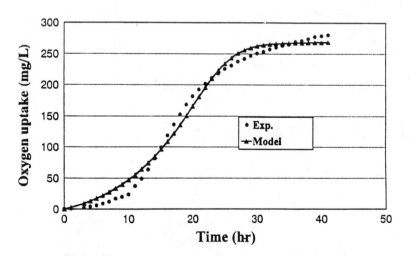

Figure 3.   Experimental data and model fit of cumulative oxygen uptake for 150 mg/L of phenanthrene using 5% soil suspension.

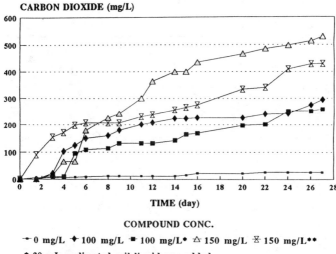

CARBON DIOXIDE (mg/L)

COMPOUND CONC.

--- 0 mg/L   ✦ 100 mg/L   ■ 100 mg/L*   △ 150 mg/L   ⌧ 150 mg/L**

  * 20 mL acclimated soil-liquid was added
 ** 12.5 g acclimated soil was added

Figure 4. Cumulative carbon dioxide evolution data for naphthalene at various concentrations in soil suspension reactors.

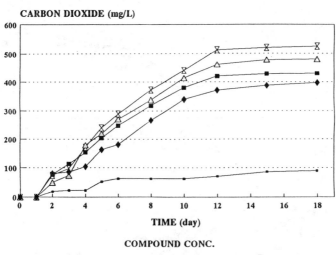

CARBON DIOXIDE (mg/L)

COMPOUND CONC.

--- 0 mg/L   ✦ 100 mg/L   ■ 100 mg/L*   △ 150 mg/L   ⌧ 150 mg/L*

  * 20 mL acclimated soil-liquid was added

Figure 5. Cumulative carbon dioxide evolution data for phenanthrene at various concentrations in soil suspension reactors.

relationship of the two reactor systems provided data on the correlation between the oxygen requirements and CO_2 generation for metabolic activity of microbiota on toxic organics in the soil slurry systems.

Studies with uniformly labelled ^{14}C PAHs showed that over 95% of the carbon dioxide evolved was produced due to PAH mineralization and less than 40 mg/L of total carbon dioxide was produced due to mineralization of soil organic matter. This showed that measurement of carbon dioxide evolution in spiked soil slurry reactors can be used to quantitate the rate of contaminant mineralization.

The carbon dioxide generation rate provides data on the mineralization rate of the contaminant. The total amount of carbon dioxide generated agrees closely with the appropriate amount of oxygen uptake required to mineralize the compound. This demonstrates that complete mineralization of the compound was achieved in the respirometric experiments.

The quantification of carbon dioxide generation rates and determination of the mass balance with oxygen uptake data is an important achievement in studying bioremediation of organics in soil systems.

Analysis of Adsorption/Desorption Data

Representative adsorption kinetic data for naphthalene and phenanthrene with model fits are shown in Figures 6 and 7, respectively. For all concentrations tested, equilibrium is attained in about 20 hours. Furthermore, in the case of naphthalene, approximately 45% of the initial amount of compound is adsorbed by the time equilibrium is attained and this percent adsorption does not vary significantly with initial concentration. However, in the case of phenanthrene, about 85% of the initial amount of compound is adsorbed until equilibrium is attained. This increase in relative percent adsorption is mainly due to partitioning into the soil organic matter which depends on the compound's octanol-water partition coefficient. The octanol-water partition coefficient for phenanthrene is about 60% higher than that for naphthalene.

The liquid concentrations for all PAHs did not vary more than 5% after 20 hours, which indicated that equilibrium was achieved. In addition, it can be concluded that since the concentration of the compound did not change after 48 hours even though the experiments were conducted for 96 hours, there was no biological degradation of the compounds under the experimental conditions.

Representative desorption data for naphthalene and phenanthrene with model fits are shown in Figures 8 and 9 respectively. Desorption is usually slower than adsorption with most of the adsorbed compound desorbing in 20 hours, although equilibrium is attained in 40 hours.

Comparing the adsorption/desorption equilibration time (approximately 20 hours) with biodegradation acclimation time (42 hours for 100 mg/L initial concentration of naphthalene and 200 hours for 150 mg/L of phenanthrene), it is clear that adsorption/desorption equilibrium is achieved much before the onset of biodegradation in soil slurry reactors. Furthermore, with the onset of biodegradation and hence oxygen uptake, it was observed that the PAHs were completely emulsified in the slurry reactors, even though insoluble particulates of PAHs were visible before the onset of oxygen uptake. This indicated that the soil microbiota was capable of synthesizing biosurfactants that solubilized the insoluble PAHs. It is expected that the presence of bioemulsifying agents in soil slurry reactors would modify the adsorption/desorption characteristics of the PAHs from the soil particles, and hence the adsorption/desorption equilibration time, measured abiotically, can no longer be compared with biodegradation time.

Analysis of the bacterial sorption data showed that equilibrium was achieved in less than 3 hours. This meant that equilibrium level of microorganisms in the water and soil phases were achieved very quickly compared to adsorption/desorption equilibration time for the PAHs or their biodegradation

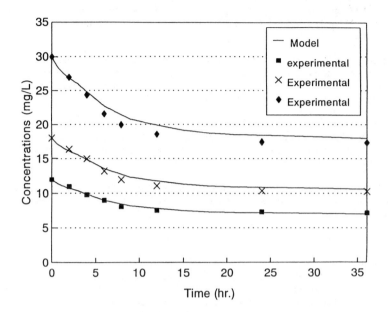

Figure 6. Abiotic adsorption experimental data with model fits for naphthalene.

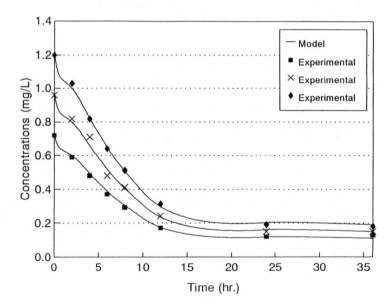

Figure 7. Abiotic adsorption experimental data with model fits for phenanthrene.

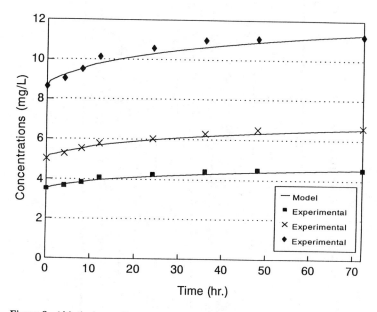

Figure 8. Abiotic desorption experimental data with model fits for naphthalene.

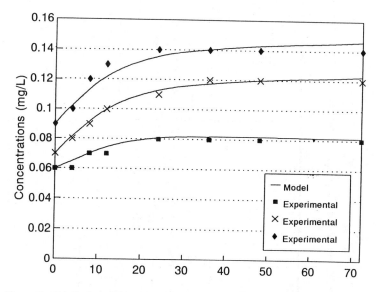

Figure 9. Abiotic desorption experimental data with model fits for phenanthrene.

Table III. Summary of kinetic and equilibria isotherm parameters for adsorption and desorption of PAHs in soil slurry systems

Compound	Adsorption Kinetics			Adsorption Isotherms		Desorption Kinetics			Desorption Isotherms	
	k_{1a}	k_{-1a}	m_a	k_a	$1/n_a$	k_{1d}	k_{-1d}	m_d	k_d	$1/n_d$
Naphthalene	0.47±0.01	0.00063±1E-5	0.474±0.001	4.60±0.01	0.900±0.001	0.187±0.001	0.00063±1E-5	0.108±0.00	4.60±0.01	0.84±0.01
Phenanthrene	0.235±0.001	0.00119±1E-5	0.0252±2E-4	24.92±0.03	0.97±0.01	0.278±0.001	0.00226±1E-5	0.478±0.00	15.4±0.02	0.67±0.01
Acenaphthylene	0.291±0.001	0.00063±2E-5	0.153±0.002	8.9±0.01	0.99±0.01	0.102±0.001	0.00120±2E-5	0.0106±1E-	7.90±0.01	0.92±0.01
Acenaphthene	0.370±0.001	0.00100±1E-5	0.141±0.001	10.1±0.01	1.01±0.01	0.144±0.001	0.00133±1E-5	0.737±0.001	10.2±0.01	1.01±0.01

Bacterial sorption isotherm parameter $K_b = 167.0±0.1$

time periods. Analysis of the bacterial sorption isotherm gave a K_b value of 167.0 which indicated that significantly higher bacterial concentrations were present in the soil phase compared with the aqueous phase. It should be noted that the ratio of bacterial concentrations present in the soil and liquid phases, as quantified by the K_b value, are present at the onset of biodegradation. The growth rates of the soil and aqueous phase microorganisms are different, and hence aqueous phase biodegradation may predominate in soil slurry reactors, even though the initial aqueous phase bacterial concentrations are much lower than the soil phase.

Table III summarizes the data on the adsorption/desorption kinetic and equilibrium isotherm parameters for the four PAH compounds. The experimental data fitted the calculated values, obtained by using the kinetic and isotherm parameters, very well, with an R^2 value greater than 0.98 for all PAH compounds. The adsorption-desorption kinetics have a direct impact on the compound's biodegradability, as shown by the mathematical model analysis of the oxygen uptake data.

CONCLUSIONS

Respirometric studies with soil slurry reactors provides valuable insight into the biodegradation kinetics of compounds in the presence of soil. It has been shown that a Monod kinetic equation in conjunction with adsorption/desorption kinetics and equilibria equations can provide reliable estimates of the Monod kinetic parameters. The adsorption/desorption kinetics and equilibria parameters were obtained by fitting abiotic adsorption/desorption data to the nonlinear rate equations and Freundlich isotherms. Experiments conducted in our laboratory have demonstrated that cumulative carbon dioxide measurement can be made for soil slurry systems. Carbon dioxide generation in soil slurry systems provides unambiguous measurements of the fate of mineralization of the compound in the presence of soil.

Reconciliation of carbon dioxide generation data with oxygen uptake information is important in determining the biokinetics of not only biotransformation reactions, but also for complete mineralization of the compound.

A protocol developed for quantitative measurement of $^{14}CO_2$ evolution rate by radiorespirometry provides confirmation of mineralization kinetics from carbon dioxide evolution studies and ensures that the net CO_2 is generated from mineralization of the compound and not due to natural soil respiration.

Further studies are planned for continuing the respirometric oxygen uptake and carbon dioxide generation measurements for other compounds. This would enable the generation of a database of Monod kinetic parameters and soil adsorption coefficients for various compounds. Eventually, this database will be used to develop predictive models using structure-activity relationships, so that biokinetics of a wide variety of compounds in soil systems can be estimated.

Further studies are also planned for measuring the biokinetics of compounds in compacted soil systems, as opposed to our current measurements in soil slurry systems. This will allow one to determine the impact of soil compaction, oxygen transfer rates, and moisture content on biodegradation.

Attempt will also be made to develop a detailed mathematical model implemented as a computer program, that will use our experimental data and model parameters to actually quantify rates of bioremediation at contaminated sites. This is significant, since at present, there is no systematic methodology for determining extent of bioremediation at contaminated sites.

REFERENCES CITED

1.   Heitkamp, M.A. and C.E. Cerniglia "Polycyclic Aromatic Hydrocarbon Degradation by a *Mycobacteriumsp.* in Microcosms Containing Sediment and Water from a Pristine Ecosystem" Applied and Environm. Microbiol. 55 (8) 1968-1979 (1989).

2.   Cerniglia, C.E. "Microbial Metabolism of Polycyclic Aromatic Hydrocarbons." Adv. Appl. Microbiol. 30, 31-71 (1984).

3.   Clover, M.P. "Studies of the Persistence of Polycyclic Aromatic Hydrocarbons in Unacclimated and Agricultural Soils" M.S. Thesis - Utah State University, Department of Civil and Environmental Engineering. (1987).

4.   Ellis, B., P. Harold and H. Kronberg "Bioremediation of a Creosote Contaminated Site" Environmental Technology 12, 447-453 (1991).

5.   Fredrickson, J.K., F.J. Brockman, D.J. Workman, S.W. Li and T.O. Stevens "Isolation and Characterization of a Subsurface Bacterium Capable of Growth on Toluene, Naphthalene, and Other Aromatic Compounds" Applied and Environ. Microbiol. 57 (3) 796-803 (1991).

6.   Gibson, D.T. and V. Subramanian. "Microbial Degradation of Aromatic Hydrocarbons." In: Gibson, D.T. (Ed.) Microbial Degradation of Organic Compounds. Dekker, New York. pp 181-252 (1984).

7.   Brodkorb, T.S. and R.L. Legge "Enhanced Biodegradation of Phenanthrene in Oil Tar-Contaminated Soils Supplemented with *Phanerochaete chrysosporium*" Applied and Environ. Microbiol. 58 (9), 3117-3121 (1992).

8.   Atlas, R.M. "Microbial Degradation of Petroleum Hydrocarbons: An Environmental Perspective" Microbiological Reviews March 45 (1) 180-209 1981.

9.   Sims, R.C. and M.R. Overcash. "Fate of Polynuclear Aromatic Compounds (PNAs) in Soil-Plant Systems" In: Residue Reviews. Residues of Pesticides and Other Contaminants in the Total Environment; Edts Gunther, F.A. and J.D. Gunther, Springer-Verlag, N.Y., Heidelberg, Berlin, 88, 1-68 (1983)

10.   Mueller, J.G., S.E. Lantz, B.O. Blattmann and P.J. Chapman "Bench-Scale Evaluation of Alternative Biological Treatment Processes for the Remediation of PCP- and Creosote-Contaminated Materials: Slurry-Phase Bioremediation. U.S.EPA, Gulf Breeze Research Laboratory, Report No. 27, (1991) 27 pp.

11.   Mueller, J.G., D.P. Middaugh, S.E. Lantz and P.J. Chapman "Biodegradation of Creosote and Pentachlorophenol in Contaminated Groundwater: Chemical and Biological Assessment" U.S.EPA, Gulf Breeze Research Laboratory, Report No. (1991) 27 pp.

12.   Nitschke, V., M. Beyer, J. Klein, D.C. Hempel. "Microbial Biodegradation of Polycyclic Aromatic Hydrocarbons" Chemische Industrie: Zeitschrift Für Die Deutsche Chemiewirtschaft, Dusseldorf, Germany, 115 (10) 53-56 (1992).

13.   Mueller, J.G., P.J. Chapman and P.H. Pritchard "Potential for the Bioremediation of Creosote-Contaminated Sites" U.S.EPA, Gulf Breeze Research Laboratory, Report No. 671 (1984) 54 pp.

14.   Means, J.C., S.G. Wood, J.J. Hassett and W.J. Banwart. "Sorption of Polynuclear Aromatic Hydrocarbons by Sediments and Soils." Environ. Sci. Technol. 14 (12) 1524-1528 (1980).

15.   Means, J.C., S.G. Wood, J.J. Hassett and W.L. Banwart. "Sorption of Amino- and Carboxy-Substituted Polynuclear Aromatic Hydrocarbons by Sediments and Soils." Environ. Sci. Technol. 16, 93-98 (1982).

16.   Wu, S-C and P.M. Gschwend. "Numerical Modeling of Sorption Kinetics of Organic Compounds to Soil and Sediment Particles." Water Resources Research 24 (8), 1373-1383 (1988).

17.   Wu, S.C. and P.M. Gachwend. "Sorption Kinetics of Hydrophobic Organic Compounds to Natural Sediments and Soils." Environ. Sci. Technol. 20 (7), 717-725 (1986).

18.   Voice, T.C. and W.J. Weber. "Sorption of Hydrophobic Compounds by Sediments, Soils and Suspended Solids - I Theory and Background." Water Res. 17 (10), 1433-1441 (1983).

19. Weber, W.J., P.M. McGinley and L.E. Katz. "A Distributed Reactivity Model for Sorption by Soils and Sediments. 1. Conceptual Basis and Equilibrium Assessments." Environ. Sci. Technol. 26 (10), 1955-1962 (1992).

20. Karickhoff, S.W. and K.R. Morris. "Sorption Dynamics of Hydrophobic Pollutants in Sediment Suspensions." Environmental Toxicol. and Chemistry 469-479 (1985)

21. Karickhoff, S.W., D.S. Brown, and T.A.Scott. "Sorption of Hydrophobic Pollutants on Neutral Sediments." Water Res. 13, 241 (1979).

22. Means, J.C., J.J. Hassett, S.G. Wood, W.L. Banwart, S. Ali and A. Khan. "Sorption Properties of Polynuclear Aromatic Hydrocarbons and Sediments: Heterocyclic and Substituted Compounds." In: Polynuclear Aromatic Hydrocarbons: Chemistry and Biological Effects. A. Bjorseth and A.J. Dennis (Eds.) Battelle Press, Columbus, Ohio, 395-404 (1980).

23. Karickhoff, S.W. "Pollutant Sorption in Environmental Systems." U.S.EPA Report. EPA-600/0-83-083, July 1983, 31 pp.

24. Brusseau, M.L., R.E. Jessup and S.C. Rao. "Nonequilbrium Sorption of Organic Chemicals: Elucidation of Rate-Limiting Processes." Environ. Sci. Technol 25 (1), 134-142 (1981)

25. Brusseau, M.L. and P.S.C. Rao. "Influence of Sorbate Structure on Nonequilibrium Sorption of Organic Compounds. Environ. Sci. Technol. 25 (8), 1501-1508 (1981).

26. Weisenfels, W.P., H.J. Kiever and J. Langhoff. "Adsorption of Polycyclic Aromatic Hydrocarbons (PAHs) by Soil Particles: Influence on Biodegradability and Biotoxicity." Applied. Microbiol. Biotechnol. 36, 689-696 (1992).

27. Mihelcic, J.R. and R.G. Luthy "Microbial Degradation of Acenaphthene and Naphthalene under Denitrification Conditions in Soil-Water Systems" Applied and Environm. Microbiol. 54 (5) 1188-1198 (1988).

28. Govind, R., C. Gao, L. Lai, X. Yan, S. Pfanstiel, and H.H. Tabak. "Development of Methodology for the determination of Bioavailability and Biodegradation Kinetics of Toxic Organic Pollutant Compounds in Soil." Paper presented at the In-Situ and On-Site Bioreclamation, 2nd International Symposium, San Diego, CA, April 5-8, 1993.

29. OECD 1981. OECD Guidelines for Testing of Chemicals Section 3, Degradation and Accumulation, Method 301C, Ready Biodegradability: Modified MITI Test (I) adopted May 12, 1981 and Method 302C Inherent Biodegradability: Modified MITI Test (II) adopted May 12, 1981, Director of Information, Organization for Economic Cooperation and Development, Paris, France.

RECEIVED June 27, 1995

Chapter 22

Uncouplers of Oxidative Phosphorylation

Modeling and Predicting Their Impact on Wastewater Treatment and the Environment

Robert W. Okey[1] and H. D. Stensel[2]

[1]Department of Civil Engineering, University of Utah,
Salt Lake City, UT 84112
[2]Environmental Engineering and Science Program,
University of Washington, Seattle, WA 98195

A study of uncoupling data from three different species has been carried out. The data were developed on activated sludge, a ciliate protozoa and plant cells grown in dispersed culture. The purpose of this work was to determine if data from several species could be used to broaden the database for all species. The procedure involved the development of a linear free energy relationship (LFER) for the quantitative structure-activity relationship (QSAR). Molecular connectivity indices and groups were employed as descriptors in the LFER. Three models were developed, one of which was rejected because of collinearity and the resulting impact on the coefficients. One of the remaining models was used to compare predicted response to that seen on *Arbacia* eggs and by organics contained in pulp mill effluents. The implication of the findings is discussed.

Many substances that do not respond in the traditional way to the biochemical oxygen demand (BOD) test are present in varying concentrations in the influent to wastewater treatment facilities. These materials may be present in such low concentrations as to have no impact on the BOD test. Such materials are usually transiently present and may go undetected even by the most sophisticated screening schemes. However, the materials may be bioconcentrated and ultimately exert an effect far beyond that expected based on influent strength data.

One group of such anthropogenic compounds, the halogenated phenols (other than pentachlorophenol), has received only limited attention despite widespread use in pesticides and organic synthesis and their substantial presence in bleached pulping effluents. Also, phenols carrying nitro groups and the halogenated and nitro-containing anilines have been subjected only to limited study. When transiently present in the influent to a wastewater treatment facility,

0097–6156/95/0607–0284$13.50/0

these materials are not degraded to any extent, nor are they toxic at the concentrations usually encountered. However, their presence can result in a substantial change in plant performance and behavior. Such changes as an increase in oxygen use without assimilation or degradation and substantial decreases in net synthesis have been seen. This behavior has been identified as the uncoupling of oxidative phosphorylation.

The uncoupling of oxidative phosphorylation occurs when the energy-yielding oxidation (catabolic) step is "uncoupled" from the formation of ATP (adenosine triphosphate) and the production of new cells (anabolic steps). Uncoupling is most frequently caused by weak lipophilic acids. In microbial systems, these materials must be refractory or slowly biodegradable to function as uncouplers. This implies that the list of materials that can uncouple in higher systems not capable of aromatic ring scission may be much larger than those impacting as microbial systems. There is some evidence that this is correct.

The actual data available from studies on activated sludge are very limited (Okey and Stensel, *1*; Beltrame *et al.*, *2,3*; Rich and Yates, *4*; Schneider, *5*). Hence, the actual inhibition database is limited to only a few materials. There are, however, data from studies on higher biological systems that could, if found to be usable, expand the list of known microbial uncouplers substantially. There is some basis for believing that such an expansion of the available database is possible as, according to Maloney (*6*), the essential processes associated with ATP production and uncoupling are similar for all types of cells.

The purpose of this work was to obtain multiple species uncoupling data and determine if a broad-based quantitative structure-activity relationship (QSAR) of reasonable significance could be developed. If successful, a much broader database would become available permitting the use, thereof, for estimating the impact of any organic on any commonly-employed test species. Also, a successful development of a broadly applicable model might be used in any setting to predict the likely impact of such materials on organisms other than those specifically under study. Data from uncoupled plant, protozoan and activated sludge systems were employed for that purpose here.

In examining the available data, the fact that "uncoupling" or "to uncouple" has at least two distinctly different meanings was kept in mind. Some workers apply the term to the Pasteur effect or the "uncoupling" of the terminal respiration sequence during the fermentative metabolism of glucose or other substrates (*7*). The extent to which the system is uncoupled is a function of the fraction of the substrate utilized anaerobically as opposed to aerobically. In the case reported by Chudoba *et al.* (*7*), the mix of microorganisms is changed by the increase in poly-P bacteria which may have a unique energy requirement associated with the capture of $PO_4^=$ and the manufacture and storage of polyphosphates. It is not clear if such energy is recoverable. It is clear that in such anaerobic-aerobic systems the synthesis is less than expected in completely-aerobic activated sludge.

However, the term as it is used in this work refers to the "uncoupling" of the energy-yielding electron transport sequence from the energy-requiring formation of adenosine triphosphate (ATP). Since production of ATP regulates cell respiration when the system is uncoupled, respiratory regulation is lost and the cell respiration rate continues to increase until intracellular reserves are exhausted (8). Symptoms of uncoupling are an increased rate of respiration, limited or no synthesis and ultimately a reduction in cell mass as reserves are exhausted.

The phenomenon of uncoupling has been recognized for some time in mammals, in yeasts and algae. The most frequently-studied uncoupler has been 2,4-dinitrophenol. However, bilirubin, azide and pentachlorophenol have all been identified as uncouplers in various biological systems (9). Servizi (10) has reported that chlorinated catechols contained in the effluent from pulp-bleaching operations can uncouple ATP formation in migrating salmon. Insecticides containing nitro-phenols are effective because of the ability to uncouple (11).

Clowes et al. (12), Clowes (13), and Krahl (14) studied the uncouplers, nitro- and halogenated phenols and related molecules, using Arbacia eggs as the test system. They noted that the location of the substituents affected uncoupling activity and that a hydroxyl group or an amine was necessary for uncoupling to occur. The uncoupling activity was evaluated by determining the increase in oxygen consumption and the decrease in cell division as a function of concentration of the uncoupling compound.

These and many other references are contained in an earlier work (1). The mechanism by which the biological system is uncoupled is discussed in detail in that same work. An overly-simplified explanation is that ATP is produced by the flow of protons across the plasma membrane. This flow occurs as a result of the existence of a pH and an electrical potential across the energized membrane. The lipid soluble biorefractory acids penetrate the membrane and move to the alkaline interior where they yield their protons, thereby destroying both the electrical and pH potential. Hence, the controlled flow of protons and the normal production of ATP is reduced in extent or eliminated altogether.

Uncoupler Database

The data of Beltrame et al. (2,3), Cajina-Quezada and Schultz (15) and Ravanel et al. (16) were employed to develop a multispecies QSAR for uncouplers. Each of these studies established a critical criterion which was sufficiently similar to the others to permit the combined use of the work. The data, to be used, has to specifically describe the uncoupling phenomenon as indicated by excess biological respiration, a decrease in the growth substrate uptake rate, a specific change in ATP synthesis or a decrease in the rate of cell synthesis.

Beltrame et al. (2,3) studied the specific impact of the uncouplers, the halogenated and nitrated phenols, on the rate of phenol metabolism. This work is similar to studies reported by Okey and Stensel (1). The critical criterion employed was the dose required to reduce the rate of metabolism of phenol, the growth substrate, by 50 percent.

The non-activated sludge studies included one using plant cells and a second using a ciliate. The plant studies used Acer (maple) cell suspensions (16). The uncoupling of Acer cells was quantified by determining the uncoupler dose required to reduce ATP production by 50 percent. Oxygen uptake and the increase in the oxygen use were also monitored. The ciliate studies utilized *Tetrahymena pyriformis* (a ciliate protozoa) (*15*). The ciliate studies evaluated the impact of the uncouplers on growth. The critical dose was that which reduced synthesis by 50 percent.

The data from each study are presented in Table I. The nomenclature employed in this work is

$$\text{Log BR (Biological Response)} = \log \frac{1}{\text{critical concentration}}$$

where the critical concentration is reported in mM/L.

Table I. Materials Used in Study of Uncouplers

Uncoupling of Activated Sludge (2)

(Reported in this study in log 1/LD50-mM/L)

No.	Compound	Log BR
1	Chlorophenol	0.090
2	3-Chlorophenol	0.280
3	4-Chlorophenol	0.258
4	2,3-Dichlorophenol	0.471
5	2,4-Dichlorophenol	0.535
6	2,5-Dichlorophenol	0.511
7	2,6-Dichlorophenol	0.398
8	3,4-Dichlorophenol	0.582
9	3,5-Dichlorophenol	0.447
10	2,3,4-Trichlorophenol	0.858
11	2,3,5-Trichlorophenol	0.947
12	2,3,6-Trichlorophenol	0.701
13	2,4,5-Trichlorophenol	0.923
14	2,4,6-Trichlorophenol	0.672
15	3,4,5-Trichlorophenol	1.003
16	2,3,4,5-Tetrachlorophenol	1.056
17	2,3,4,6-Tetrachlorophenol	0.758
18	2,3,5,6-Tetrachlorophenol	0.719
19	Pentachlorophenol	1.058
20	2-Nitrophenol	-0.027
21	3-Nitrophenol	0.085
22	4-Nitrophenol	0.107
23	2,5-Dinitrophenol	1.015
24	2,6-Dinitrophenol	0.558
25	3,4-Dinitrophenol	1.244

Table I. *Continued*

Uncoupling of Activated Sludge (3)

No.	Compound	Log BR
1	2-Nitrophenol	-0.197
2	3-Nitrophenol	0.075
3	4-Nitrophenol	0.024
4	2,5-Dinitrophenol	0.759
5	2,6-Dinitrophenol	-0.628
6	3,4-Dinitrophenol	0.151
7	2-Chloro-4-nitrophenol	-0.150
8	3-Chloro-4-nitrophenol	0.309
9	4-Chloro-2-nitrophenol	0.025
10	4-Chloro-3-nitrophenol	0.522
11	2,4-Dichloro-6-nitrophenol	0.123
12	2,6-Dichloro-4-nitrophenol	-0.025

Uncoupling of *Acer* (Maple) Cells (16)

(Reported in this study in log 1/LD50-mM/L)

No.	Compound	Log BR
1	2-Chlorophenol	0.001
2	3-Chlorophenol	0.001
3	4-Chlorophenol	0.155
4	2,3-Dichlorophenol	0.699
5	2,4-Dichlorophenol	0.824
6	2,5-Dichlorophenol	1.155
7	2,6-Dichlorophenol	0.155
8	3,4-Dichlorophenol	1.301
9	3,5-Dichlorophenol	1.523
10	2,3,6-Trichlorophenol	1.602
11	2,4,5-Trichlorophenol	2.000
12	2,4,6-Trichlorophenol	1.000
13	Pentachlorophenol	2.301
14	4-Chloro-2-methylphenol	0.824
15	4-Chloro-3-methylphenol	0.523
16	4-Chloro-2,3-dimethylphenol	0.024
17	4-Chloro-3,5-dimethylphenol	1.155
18	4-Chloro-2-allylphenol	1.155
19	4-Chloro-2-isopropyl-5-methylphenol	1.155
20	2-Chloro-6-nitrophenol	0.398
21	2,4-Dichloro-6-nitrophenol	1.523
22	2-Chloro-4,6-dinitrophenol	0.097

Table I. *Continued*

Uncoupling of *Tetrahymena pyriformis* (15)

(Reported in this study in log 1/LD50-mM/L)

No.	Compound	Log BR
1	4-Phenylazophenol	1.655
2	2,4-Dinitrophenol	1.096
3	2,5-Dinitrophenol	0.929
4	2,6-Dinitrophenol	0.573
5	2,4-Dinitro-1-naphthol	1.585
6	2,6-Dinitro-4-methylphenol	1.230
7	4,6-Dinitro-2-methylphenol	1.329
8	2,6-Dibromo-4-nitrophenol	1.357
9	2,4-Dichloro-6-nitrophenol	1.750
10	2,6-Diiodo-4-nitrophenol	1.812
11	2,3,4,5-Tetrachlorophenol	2.004
12	2,3,5,6-Tetrachlorophenol	2.219
13	2,3,5,6-Tetrafluorophenol	1.167
14	3,4,5,6-Tetrabromo-2-methylphenol	2.573
15	Pentabromophenol	2.664
16	Pentachlorophenol	2.568
17	Pentafluorophenol	1.631
18	4-Phenylazoaniline	1.421
19	2,4-Dinitroaniline	0.716
20	2,6-Dinitroaniline	0.941
21	3,5-Dinitroaniline	0.941
22	2,4-Dibromo-6-nitroaniline	1.170
23	2,4-Dichloro-6-nitroaniline	1.170
24	4,5-Dichloro-2-nitroaniline	1.714
25	4,5-Difluoro-2-nitroaniline	0.764
26	2,3,4,5-Tetrachloroaniline	1.956
27	2,3,5,6-Tetrachloroaniline	1.762

Earlier work (*29*) on the data sets employed here demonstrated the ability to predict across species boundaries but a unified model was not developed. The data were analyzed individually and simple models obtained. These are reported below.

For the activated sludge study the Beltrame *et al.* data base produces the following LFER.

$$\text{Log BR} = 0.143 \ (^{0}X^{v}) + 4.021 \ (^{7}X_{CH}) - 1.137$$

$$n = 30, S_e = 0.129, r^2 = 0.880, r^2(df) = 0.871, F = 99.01.$$

The data of Cajuna-Quezaela and Schultz on the uncoupling of the ciliate *Tetrahymena pyriformis* produced a somewhat different model, primarily because all of the uncouplers were not phenols.
The model is as follows:

$$\log BR = 0.144\ (^\circ X^v) + 0.196\ (OH) + 8.888\ (^7X_{CH}) - 1.492$$

$$n = 25, S_e = 0.241, r^2 = 0.835, r^2(df) = 0.811, F = 35.33.$$

The model developed on *Acer* cells is as follows:

$$\log BR = 0.235\ (^\circ X^v) + 12.089\ (^7X_{CH}) + 0.295\ (HAL)$$

$$- 0.265\ (ORTHO) - 3.231$$

$$n = 19, S_e = 0.269, r^2 = 0.868, r^2(df) = 0.831, F = 23.06.$$

The outliers dropped during these studies are as shown in Table II.
    These models demonstrated that it is possible to explain between 80 and 85% of the variance using two to four variables. The intent in this work is to demonstrate only that it is possible to build a model that explains 70 to 80% of variance using data from the three species utilized here. The impact on activated sludge has been demonstrated and reported in Okey and Stensel (1).

Research Methods

Introduction. The key consideration in the development of a QSAR (after data sufficiency) is the method by which the variables are selected. The linear free energy relationship (LFER) that is used for the QSAR must be robust, free from collinearity and must offer proof that the best possible set of descriptor variables has been included. This latter requirement is often the most difficult to assure because of the large number of descriptors available.
    The relevant issues concerning the construction of an LFER have been dealt with in detail in two other publications (17, 18) and are summarized only briefly here. The linear free energy relationship is based on the concept that a modest change in molecular structure, such as the addition of a substituent, results in a linear change in the activation energy of a rate-limiting step. The change in reaction rate brought about by the change in activation energy is described or modeled by the LFER. The LFER is a combination of constants and molecular descriptors which is used to describe molecular structure.
    The effect of substituents can be categorized as either electronic, steric, or both. Electronic effects occur because of an alteration in the "π" or "σ" electron density at various sites in the molecule. This effect derives from the ability of substituents to contribute or withdraw electrons. This effect is very pronounced

Table II. Outliers Dropped during Model Development

Test System	Compound	First Regression Value-Log BR mM/L	Observed Value-Log BR mM/L
Activated Sludge	2,5-Dinitrophenol	0.388	1.015
	2,5 Dinitrophenol	0.328	0.759
	2,6-Dinitrophenol	0.070	-0.628
	3,4-Dinitrophenol	0.589	1.244
	2-Chloro-4-nitrophenol	0.417	-0.150
	4-Chloro-2-nitrophenol	0.417	0.025
	2,4-Dichloro-6-nitrophenol	0.657	0.123
Ciliate	2,5-Dinitro-1-naphthol	1.585	2.144
	Pentachlorophenol	2.568	1.960
	2,6-Dinitro-4-methylaniline	0.100	0.839
	2,6-Dibromo-4-nitroaniline	0.100	0.839
	2,6-Dichloro-4-nitroaniline	0.100	1.031
Acer (maple) cells	4-Chloro-2-allylphenol	0.474	1.155
	2,4-Dichloro-6-nitrophenol	0.887	1.523
	2,3,4-Trichlorophenol	1.296	0.602

when the substituent is an electronegative atom such as nitrogen, oxygen, or a halogen. Steric effects derive from the physical blocking of a reaction center or may be the result of a poor fit with the enzyme created by the substituent.

The LFERs are of the form:

$$\text{Log } K = AK_1 + BX_2 + CX_3 + DX_4 + E \qquad (1)$$

where:

K	=	the reaction rate,
A, B, C, D, & E	=	regression constants, and
X_1-X_4	=	descriptor variables.

The descriptor variables and the validity of their selection determines, in large measure, the extent to which the phenomenon is properly modeled. A later section is devoted entirely to the process of variable selection.

Statistical Methods

Required Software and Hardware. The development of the quantitative structure-activity relationship (QSAR) is a manageable task only with the use of a computer and the appropriate software. The needs are:
1. A computer system capable of rapidly processing several thousand data entries.
2. A software package capable of carrying out regressions with up to several hundred cases and as many as sixty variables.

The computers employed were based on the 386 chip in PC clones. The statistics software employed were Microstat II (*19*) and SPSS (*20*). Both software packages were capable of regression, ANOVA and residual analyses. Microstat II does not contain a sub-routine for collinearity diagnostics; however, this menu-driven software was easier to use for much of the work. SPSS was used to obtain the critical collinearity diagnostics, after Belsley *et al.* (*21*), and specific residual information.

Graph III (*22*) was employed to obtain the molecular connectivity indices. This software computes the indices up to the sixth or seventh order. The input required is the bond and valence information. The graphs were produced using SigmaPlot 5.0 (*23*).

Control of Collinearity. Collinearity occurs when two or more variables contribute the same property to the regression. The correlation matrix can be relied upon to identify the problem if only two variables are involved; however, often three or more variables are involved and then the correlation matrix will not sufficiently quantify the problem to alert the researcher. The procedure of Belsley *et al.* (*21*) was used here. This technique is based on the value of the condition index which is the square root of the ratio of the maximum eigenvalue divided by the individual eigenvalues. According to Belsley *et al.* (*21*), when the condition index is greater than 30 and more than one variable has a high regression coefficient variance (>0.50), the collinearity is sufficiently degrading so that steps must be taken to correct the problem. This is such a significant problem that Chatterjee and Price (*24*) indicate that many studies using regression contained in the literature are improperly done due to collinearity.

Elimination of Outliers. A procedure called Cook's distance was employed for identifying influential outliers. Cook's distance is a measure of the extent of influence of a given case on the regression coefficients. This method is described in most texts on regression procedures. Cook[25] originally reported the technique in 1977. The procedure is, of course, extremely difficult to perform without a computer.

One way of interpreting Cook's distance is that it is the standardized weighted distance between the regression coefficient obtained with the full data set and that obtained by deleting the "i[th]" observation. Cook's distance can be determined as a part of the SPSS regression procedure dealing with residuals.

Tests of Variable Significance. There are several standard statistical tests that indicate the relative strength or robustness of the entire model and, more importantly, of individual variables. Two were used here. The two tests are the "t" statistic and the "p_x." The latter is a statement indicating that the probability that a variable is "correct" is >0.99, or that it is not correct is <0.01 as related to the acceptance of "null" hypothesis. The "t" statistic is related to the standard error of the regression coefficient; the larger the "t," the smaller the standard error. The "t" value is tested by comparing the calculated value with tabulated critical values. The "t" statistic is also a measure of the goodness of fit.

All regression software packages report "t" and the corresponding "p_x" value. The "t" values are given for various values of "p_x" at several significance levels. All statistics packages generate information concerning a specific regression which includes the standard error, the "t" statistic, the probability, the regression coefficient and the "ß" coefficient for each variable.

Model Quality and Validation. The quality of the model, in the last analysis, is based on its ability to accurately predict the properties represented by the dependent variable. However, the quality of the regression can be confirmed, in part, by certain aspects of the statistics employed. Some of the important parameters, or system properties, required for robust models are the following.

1. There should be five or more cases for each independent variable (*26*).
2. The variables should be selected from groups based on the nature of the information contributed to the regression[27].
3. The regression should produce a correlation coefficient "r^2" which indicates that the majority of the variance is accounted for by the equation and the actual value of "r" is greater than tabulated critical values.
4. The model should produce an "F" ratio showing that the majority of the variance is explained by the model. The "F" ratio is the ratio of the variance accounted for in the model and the unexplained variance. This is a normal computer output.

It is assumed that, in making the statistical study, the data are normally distributed (which was confirmed) and that the index and rate values used in calibration are independent of each other. To the extent possible, these assumptions are confirmed as part of the work.

Variable Selection and Validation. There are many types of descriptors that can be used in an LFER. These are described more completely elsewhere (*1, 17, 18*). The traditional equations for toxicity prediction have been based primarily on log P augmented, in some instances, by the appropriate Hammett's constants and, more recently, by a selected molecular connectivity index (*28*). Log P has been shown to be a useful descriptor, in many cases. Log P is a nonspecific descriptor reflecting the contrasting rates of penetration through lipophilic or hydrophilic layers usually based on partitioning between octanol and water. Hammett's constants are directional in character (para and meta); therefore, knowledge of the reaction mechanism and active sites is required for their intelligent use (*17*).

After a review of the extensive literature (29), the use of groups or fragments combined with molecular connectivity indices was decided upon for the earlier work. This approach has been used here as well. Groups are identifiable molecular components or fragments. Hansch and Leo (30) have carried out extensive studies on the use of fragments (groups) for estimating Log P. Groups could indeed be used alone; however, when used alone, a descriptor is necessary for a group in each possible setting. For example, an aliphatic hydroxyl (primary, secondary or tertiary) or an aromatic hydroxyl. This problem increases the number of groups to the point where an enormous database is required for any study. This is the reason for use in combination with molecular connectivity indices. The latter serves to describe the conditions, both electronic and steric, of the precise setting in a small number of terms.

The molecular connectivity index is a method of encoding the steric and electronic properties of a molecule and which offers, perhaps, the most complete and effective technique for doing so available. The concept is described completely in the Appendix to an earlier paper on QSBR development (17). The approach is based on the Randic (31) branching index. The Randic algorithm for the branching index is $\Sigma(\delta_1\delta_1)^{-0.5}$, where the '$\delta$' values are the number of adjoining carbon atoms. In the original Randic work, all non-hydrogen atoms were treated as carbon atoms; that is, 2-methyl butane had the same index as secondary butanol. Counts of adjacent atoms along paths of various lengths constitute the 'δ' values.

Kier and Hall (28), in modifying the concept, added a second set of indices which incorporates information about lone electron pairs and 'π' electrons or valence information. A complete list of the valence 'δ' values is available from several sources including Kier and Hall (28). For example, the valence 'δ' for oxygen is 5 in the hydroxyl form and 6 in the keto or ether form. For the hydroxyl, it is 5 based on two lone pairs and one adjacent carbon. In the keto or ether group, it is 6 for two lone pairs and two adjacent carbon atoms. The 'δ' values for the other electro-negative atoms are obtained in a similar fashion. The application of the concept is shown below.

For example, s-butanol

$$
\begin{array}{c}
\text{OH} \\
| \\
\text{CH}_3 - \text{CH}_2 - \text{C} - \text{CH}_3 \\
| \\
\text{H}
\end{array}
$$

The first order (1 bond length) simple (branching) index:

$$
= \frac{1}{\sqrt{1\cdot2}} + \frac{1}{\sqrt{2\cdot3}} + \frac{1}{\sqrt{1\cdot3}} + \frac{1}{\sqrt{1\cdot3}}
$$

$$
= 0.707 + 0.408 + 0.577 + 0.577
$$

$$
= \underline{2.270}
$$

The first order valence index:

$$= \frac{1}{\sqrt{1 \cdot 2}} + \frac{1}{\sqrt{2 \cdot 3}} + \frac{1}{\sqrt{5 \cdot 3}} + \frac{1}{\sqrt{1 \cdot 3}}$$

$$= 0.707 + 0.408 + 0.258 + 0.577$$

$$= \underline{1.951}$$

Since both indices contain "σ" electron information, the difference between the simple index and the valence index provides specific knowledge on lone pairs here and "π" electrons in structures containing "π" bonds.

Kier and Hall (28), in addition to contributing the valence concept, added indices which provide more information concerning the steric nature of a molecule.

These included the cluster:

the path cluster:

and the chain:

Finally, they adopted the Greek letter 'X' (chi) as the symbol; hence, the following general nomenclature was then put into use:
1. ^0X and $^0X^v$ (a count of adjacent atoms);
2. ^1X and $^1X^v$ (adjacent atoms at the end of one bond length path);
3. ^2X and $^0X^v$ (same as above except along two bond lengths path);
4. ^mX_p and $^mX_p^v$ (same as above but along three or more bond lengths path);
5. ^mX_c and $^mX_c^v$ (cluster indices);
6. $^mX_{pc}$ and $^{m4}X_{pc}^v$ (path cluster indices);
7. $^mX_{ch}$ or X_{ch}^v (chain indices).

Having settled on the use of the indices, the remaining task was to choose the correct indices from all possible for each molecule or groups of molecules. Table III is a printout for benzene from Graph III (22) which shows values and the number of indices possible up to the seventh order.

Table III. Printout of Indices for Benzene

SIMPLE MOLECULAR CONNECTIVITY INDICES

P	4.243	3.000	2.121	1.500	1.061	.750	.000	.000
C				.000	.000	.000	.000	.000
PC					.000	.000	.000	.000
CH				.000	.000	.000	.125	.000

VALENCE MOLECULAR CONNECTIVITY INDICES

P	3.464	2.000	1.155	.667	.385	.222	.000	.000
C				.000	.000	.000	.000	.000
PC					.000	.000	.000	.000
CH				.000	.000	.000	.037	.000

The method employed to select the connectivity indices for the model is shown below. Suggestions contained in James and McCulloch (27) for grouping variables with similar properties were used.

1. Descriptors supplying both electronic and size information were grouped together; for example, $^{0}X^{v}$, ^{0}X-$^{0}X^{v}$, $^{1}X^{v}$, ^{1}X-$^{1}X^{v}$.
2. Descriptors supplying size information were grouped together; for example, ^{0}X, ^{1}X, ^{2}X, $^{2}X^{v}$, ^{3}X, $^{3}X^{v}$.
3. Descriptors defining molecular complexity were grouped together; for example, $^{3}X_{c}$, $^{3}X_{c}^{v}$, $^{4}X_{pc},...^{5}X_{c}$, $^{6}X_{pc}^{v}$.
4. Descriptors for chain structures or for perimeter length were grouped together; for example, $^{5}X_{ch}$, $^{5}X_{ch}^{v}$, ... $^{7}X_{ch}^{v}$.

A variable from each category was tested in the regression, one at a time, and was maintained in the regression until all outliers were eliminated. This is the most significant single effort associated with this type of work. It must be done carefully and completely, and each group of variables that appears to provide a good regression must be checked for degrading collinearity.

In detail, the procedure employed was as follows:

1. All cases were included in the first regression.
2. After the first regression, the outstanding outlier was removed, based on Cook's distance.
3. The regression was repeated and the largest residual outlier removed.
4. The regression and elimination process was repeated until all significant outliers had been removed, based on Cook's distance and the residual values themselves (Residuals with values ≥ 2.5-3.0 times the S_{resid} were usually removed during this process.)

At this point, the significance based on the "t" statistic and "p_x" was used as a guide for retention or replacement of an index variable. Generally, a "p_x" <0.05 was used to retain. If the "p_x" was greater than 0.05, but still relevant at "p_x" <0.1, the variables were replaced but tested later with a more refined variable set.

For example, "p_x" for $^0X^v$ might have been 0.35. In which case, it would be replaced and only tested much later as part of a confirming series of regressions. It would be replaced in the next series by $^1X^v$, still employing all the groups and "dummy" variables. This process was repeated until a full set of robust variables was found.

After completion of this part of the process, the variables showing some significance but replaced initially were cross-checked against variables with high significance. In this way, the likelihood of overlooking a significant variable or variable set was reduced. Also, after the completion of each set of regressions, the relevance of other variables in each category was checked. For example, if $^6X^v_{ch}$ showed substantial relevance, $^5X^v_{ch}$, $^5X_{ch}$, $^6X_{ch}$, $^7X_{ch}$, and $^6X^v_{ch}$ were checked for significance.

The process of elimination was continued until the descriptor combinations with the highest "t" statistic or lowest "p_x" values were finally selected. The "p_x" value, the "t" statistic and the "ß" values determine the significance and robustness of each variable.

All descriptors up to the 7th order were checked independently for relevance. However, the list rapidly narrowed to include the following:

Groups: (always in regression)
1. HAL (Halogens)
2. NO$_2$ (Nitro)
3. OH (Hydroxyl)
4. NH$_2$ (Amines)

Connectivity Indices
1. $^0X^v$
2. $^7X_{ch}$
3. ^3X_c

"Dummy" Variables
1. ORTHO

This "dummy" variable was added as a surrogate or adjunct of ^3X_c.

Further and very important guidelines for the selection of variables were provided by the authors contributing the database for this study (2, 3, 15, 16). These studies detail the nature of the descriptor each worker used. Also, their literature sources provided additional information. These sources indicated that the descriptors should include terms for molecular charge, molecular volume, molecular perimeter length, steric hindrance at the ortho site and the specific groups in the molecules studied.

This suggested that the following descriptors would encode the correct information.

A. Groups
1. HAL (Halogens)
2. NO$_2$ (Nitro)

  3. OH (Hydroxyl)
  4. NH_2 (Amines)
B. Molecular Connectivity Indices
  1. $^6X_{ch}$ or $^6X_{ch}^v$ (perimeter length measure)
  2. $^7X_{ch}$ or $^7X_{ch}^v$ (perimeter length measure)
  3. ^0X to ^4X (molecular size and volume)
  4. $^0X^v$ to $^3X_p^v$ (molecular size, volume and charge)
  5. ^3X_c or $^3X_c^v$ (complexity branching measure)
  6. ^5X_c or $^5X_c^v$ (complexity branching measure)
The variables selected are contained in this list.

RESULTS AND DISCUSSION

The work, following the identification of the primary variables, led to the development of three regressions of satisfactory quality. The work involved the identification and elimination of nine outliers (shown in Table IV), three from each contributor to the database. Cook's distance indicated that outliers with values >0.03-0.04 were suspect and subject to removal. The outliers removed contained a large number of diortho-substituted materials but were not unusual or biased in other ways.

Table IV. Outliers Removed and First Regression Findings

Compound	Source	Observed	Predicted
2,6-Dinitrophenol (4)	(1)	-0.628	0.548
2-Chloro-4-nitrophenol (4)	(1)	-0.150	0.626
2,4-Dichloro-6-nitrophenol (4)	(1)	0.123	1.038
2,4,5-Trichlorophenol	(2)	2.000	1.085
Pentachlorophenol	(2)	2.301	1.613
2-Chloro-4,6-dinitrophenol	(2)	0.097	0.923
2,3,5,6-Tetrachlorophenol	(3)	2.219	1.380
Pentachlorophenol (4)	(3)	2.568	1.613
Pentafluorophenol	(3)	1.631	0.670

(1)   Beltrame *et al.* (*2, 3*)
(2)   Ravanel *et al.* (*16*)
(3)   Cajina-Quezada and Schultz (*15*)
(4)   Removed in Individual Species Models as outliers.

Several variable combinations were tried and these appeared to be satisfactory, however, two contained sufficiently collinearity to impact on the repression coefficient.

The equation for Log BR using the remaining model is as follows:

$$\text{Log BR} = 0.237\ (^{0}X^{v}) + 5.783\ (^{7}X_{ch}) + 0.163\ (NO_2)$$

$$+ 0.220\ (NH_2) - 1.944$$

$$n = 77,\ S_c = 0.329,\ r^2 = 0.724,\ r^2(df) = 0.709$$

This model has a satisfactory level of significance and a condition index of 16.675 and four variances greater than 0.5.

This model is satisfactory and is sufficiently free from degrading collinearity to produce valid regression coefficients. The observed and predicted values using this equation are shown in Figure 1.

The second and third models with "r^2" values of 0.755 and 0.724 account for over 70 percent of the variance in the data. A question always remains as to what this really means. What is the "normal" variance that would be expected if the study was done with no errors or variations other than the natural differences between test animals in the several sets used for study?

The observed and expected values vary in a similar fashion to the extent seen in a study of degradation kinetics where 155 items of rate data were evaluated for scatter *(18)*. The rate data were taken from Pitter and Chudoba *(32)*. With a mean log rate value of 1.57, the standard deviation was 0.32. That is, for example, with a characteristic rate for a given substance of 40 mg COD/g MLSS·hr (the authors' units), one standard deviation, plus and minus, ranges from 17.8 to 77.6 mg COD/g MLSS·hr. The actual range was larger but this provides some insight into the nature and extent of the variations. These variations produce "r^2" values of 0.70-0.80.

There is very little useful information in the literature with which one can evaluate the magnitude of expected variations or experimental "noise." These data provide some insight into what might be termed "normal" variation. The scatter of the predicted and observed log BR values is of the same order as the scatter of the rate data. Even though the studies were for different purposes, there is every reason to believe that both test systems could be expected to have similar variation assuming the test work was carried out without error or bias.

Significance of Findings

General Comments. It must be reemphasized here that the purpose of this work was not to develop a predictive equation, but alternatively to use all available valid data to determine if the bulk of the variance could be accounted for in a single model. If a predictive model had been desired, some of the data base would have been used for calibration and the remainder for validation. However, such was not the case. The point of this work was to determine if a

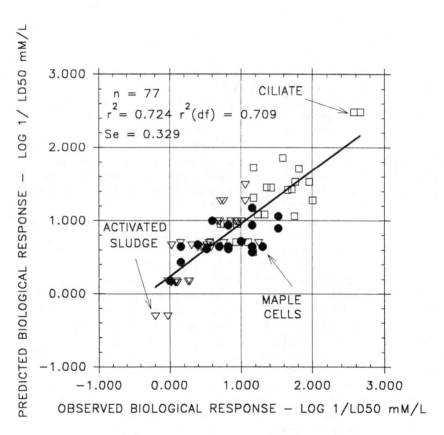

Figure 1.   Observed vs predicted biological responses to uncouplers.

reasonable correlation could be obtained combining the limited data bases for three species. Such has been demonstrated as 70% plus explanation of variance with an F ratio approaching 50 is a robust regression assuming no degrading collinearity.

It is interesting that the significant variables combined the following properties at the relevance shown. $^0X^v$ at $p < 10^{-11}$ is a measure of size and charge. $^7X_{ch}$ at $p < 10^{-4}$ is a measure of perimeter length. All substituted 6 member rings will have a $^7X_{CH}$ value. NO_2 @ $P < 10^{-2}$ is a strong electron withdrawing substituent that also enhances biorefractivity and NH_2 @ $P < 10^{-1}$ must be a proton carrier in the anilines.

So one proton carrier and one electron gathering substituent became significant variables as opposed to a halogen (another electron gathering substituent) or the hydroxyl group (the other significant proton carrier). Unfortunately, the actual qualified data base is very small. Therefore, a qualitative validation of the model must come from sources of a marginal nature. What is sought are data relating the concentration of the uncoupler, to excess oxygen use or a reduction in net synthesis, both symptoms have been clearly identified in Okey and Stensel (1) as being uniquely associated with the uncoupling of oxidative phosphorylation.

Uncoupling of *Arbacia* Eggs. Data from Clowes *et al.* (12) and Clowes and Krahl (13,14) are used here in a qualitative comparison of computed uncoupling strength and the excess respiration of *Arbacia* eggs. The data on *Arbacia* eggs are not quantifiable because the egg mass used is not included in the description of the work. The ratio of maximum uncoupled respiration and the calculated value of Log BR using model #3 (Eq. 4) are shown in Table V. The predicted Log BR values generally follow the data derived from these early studies.

Phoenix - Apparent Uncoupling of Activated Sludge. During recent work at the 23rd Avenue Treatment Plant in Phoenix, behavior of the activated sludge system similar to uncoupling was noted. (Typical data during plant upsets are shown in Figures 2, 3, 4 and 5.) The net synthesis is seen to go to zero on several occasions, and sometimes for 2-3 days. Longer periods have been seen. The COD removed may or may not change during the period of upset. Yields of 0.25-0.35 lb $MLSS/lb \ COD_R$ are considered normal. The oxygen requirement increases dramatically and SRT control becomes virtually impossible, as can be deduced from the figures. These are all the classical symptoms of uncoupling. However, based on the available data, no organic uncoupler has been detected in concentrations high enough to produce the effect seen in these figures. Recent reports (33, 34, 35) suggest that one aspect of the toxic action of certain metals is that of metabolic uncoupling. The mechanism by which this occurs is unclear and must be different from the action of weak lipophilic organic acids. However, if metals can uncouple, plant upsets with symptoms of uncoupling may occur elsewhere but go unidentified as uncoupling. Nonetheless, the system behavior in which synthesis drops to zero for several days in the presence of substantial substrate, which is removed, is a strong indicator of the presence of an uncoupler.

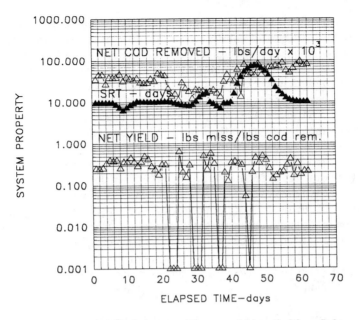

Figure 2.   Phoenix 23rd Avenue Plant solids yield – July and August, 1990.

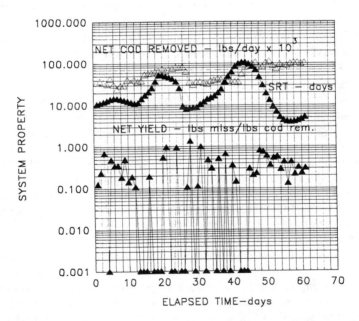

Figure 3.   Phoenix 23rd Avenue Plant solids yield – November and December, 1990.

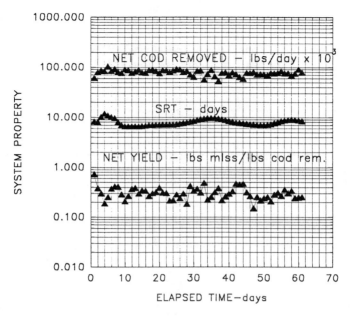

Figure 4.   Phoenix 23rd Avenue Plant solids yield - June and July, 1991.

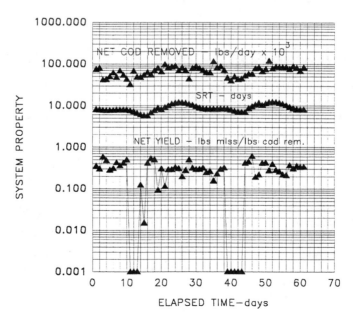

Figure 5.   Phoenix 23rd Avenue Plant solids yield - August and September, 1991.

Table V. Comparisons of Model predictions with Excess Oxygen Use by *Arbacia* Eggs (*12, 13, 14*)

Substance	Excess Oxygen Use (% Above Control)	Predicted Log BR (Equation #4)
o-Cresol indophenol	70	1.828
2,4-Dinitrothymol	Toxic - No respiration above control.	2.517 (1)
2,4-Dinitro-α-naphthol	25	1.484 (1)
4,6-Dibromocarvacrol	0	0.895 (1)
4,6-Dinitrocarvacrol	240	1.478 (1)
2,4,6-Triiodophenol	70	1.971 (1)
2,4,6-Tribromophenol	130	1.588 (1)
4-Nitrosophenol	30	0.133
4-Nitrobenzylalcohol	0	0.195
4-Nitroanisole	0	0.259 (2)
Pyocyanine	200	2.448
Tetramethyl-*p*-phenylenediamine	260	1.003 (2)
Dimethyl-*p*-phenylene	125	0.755

(1) Diortho substitution
(2) No obvious available proton

Sediments as Hazardous Waste. Because of the partitioning characteristics of most persistent organics, most river and many lake sediments in this country and elsewhere in the industrial world are heavily contaminated. Solving the problem of polluted sediments may represent the most serious remediation challenge of the next century. This problem is exacerbated by the normal process of transport to the water column, ingestion, bioconcentration, death and release back into the cycle. This cycle tends to take a great quantity of the materials from an area of possible degradation (anaerobic or anoxic sediments) into the biota that have no enzyme systems capable of degrading aromatic structures or other organics stablized by substitution.

Space limitations prevent a more complete discussion of the problem; however, one single well-documented example will serve to illustrate the nature, extent and the possible consequences of the problem.

Several Swedish workers have studied the accumulation of halogenated materials derived from pulping. Neilsen *et al.* (*36*) have documented the widespread distribution, the likely fate and comparative persistence of many of these materials in the Baltic Sea. The studies of Wulff and Rahm (*37*) and Wulff *et al.* (*38*) on sediments in the Baltic and Bothnian Seas have produced the startling report that over 50 percent of the halogenated organics discharged as effluents since the 1940s still exist on the bottom of these seas. The findings of Jonsson *et al.* (*39*) largely confirm this result. The total organically-bound chlorine from both the Swedish and the Finnish mills is estimated at 28,000 ton/yr, the bulk of which is discharged to the Bothnian Sea.

The overall rate of degradation in the sediments tested in microcosms is very small, of the order of 3.0 to 5.0 percent/yr, or less than the rate of deposit. The rate of movement or loss through the Kattegat into the North Sea is about 5.0 percent of the accumulated contaminants and is of the order of the amount currently produced and discharged. A gradual decrease is thereby predicted with the accumulated load decreased by 50 percent in the year 2040.

The chemical nature of this refractory material is relevant to this study and to the study of hazardous wastes, in general. The practice of pulp bleaching with chlorine produces many biorefractory compounds, many of which are uncouplers. A partial list of identified chlorinated organics is presented in Table VI. The starred compounds in Table VI have available protons and are suspected uncouplers. The estimated biological response using the crude model developed here for each of these suspected uncouplers is shown in brackets next to the material. This indicates that these materials may be strong uncouplers.

The overall ecological aspect of this accumulation is not simply related to the accumulation of materials on the bottom of the sea but to the fact that exchange with the water column and bioconcentration by plankton and larger members of the food chain will occur. The inevitable movement of the organisms, their death and the release of the refractories for repeated exchange and bioconcentration points to the need to study all aspects, sediments, water column and biota to ensure a complete understanding of the nature and magnitude of the problem. These sediments can hold a sufficient quantity of these materials to reduce the productivity of this great body of slightly brackish water.

Table VI. Halogenated Compounds Found in Effluents from Bleached Kraft Pulp Mills (40)

1,2-Dichlorobenzene
1,3-Dichlorobenzene
1,4-Dichlorobenzene
1,2,4-Trichlorobenzene
Hexachlorobenzene

Bis(2-chloroethoxy)methane
Bis(2-chloroethyl)ether
Bis(2-chloroisopropyl)ether
4-Chlorophenylphenyl ether
2-Chloronaphthalene
3,3′-Dichlorobenzidine

2-Chlorophenol*(0.181)
3-Chlorophenol*(0.181)
4-Chlorophenol*(0.181)

Hexachlorobutadiene
Hexachlorocyclopentadiene
Hexachlormethane

1,1-Dichloromethylsulfone
1,1,3-Trichloro-
 dimethylsulfone
Tetrachlorothiophene*(0.445)

6-Chlorovanillin*(0.921)
5,6-Dichlorovanillin*(1.247)
Trichlorosyringol*(Structure uncertain)
4-Bromophenylether

4-Chlorocatechol*(0.534)
3,5- Dichlorocatechol*(0.934)
4,5-Dichlorocatechol*(0.915)
3,4,5-Trichlorocatechol*(1.209)
3,4,5,6-Tetrachlorocatechol*(1.409)

2,3-Dichlorophenol*(0.646)
2,4-Dichlorophenol*(0.646)
2,6-Dichlorophenol*(0.646)
3,4-Dichlorophenol*(0.646)
3,5-Dichlorophenol*(0.646)
2,3,5-Trichlorophenol*(1.000)
2,3,6-Trichlorophenol*(1.000)
2,4,6-Trichlorophenol*(1.000)
3,4,5-Trichlorophenol*(1.000)
2,3,4,5-Tetrachlorophenol*(1.279)
2,3,5,6-Tetrachlorophenol*(1.380)
Pentachlorophenol*(1.613)

4-Chlorogquaiacol*(-0.113)
Dichloroguaiacol*(1.057)
4,6-Dichloroguaiacol*(1.057)
3,4,5-Trichloroguaiacol*(1.370)
4,5,6-Trichloroguaiacol*(1.370)
3,4,5,6-Tetrachloroguaiacol*(1.411)

*Compounds with available protons and predicted log BR values

Other Uncouplers. Other uncouplers have been identified and reported upon. Carr *et al.* (*41*) report on the uncoupling of rat liver mitochondria by triclabendazole. Interference with photosynthetic electron transport (PET) by halogenated 4-hydroxypyridine derivatives has been reported (*42*). Hydroxycinnamic acid has been shown to uncouple pea chloroplasts by Muzafarov and Zolotareva (*43*). The antibiotic sporaviridins uncouples phosphorylation in rat liver mitochondria *44*.

Some commonly-encountered natural antibacterials have been shown to uncouple oxidative phosphorylation. A bacteriocin produced by lactic acid bacteria is one example (*45*). Also, nisin produced by the same group of organisms has strong uncoupler properties (*46*).

These high molecular weight substances probably uncouple by a different mechanism from the phenols, anilines and pyridines. This has not been elucidated nor has the mechanism by which the metals uncouple been described. Clearly, more work is needed on the identification of the precise mechanism.

Related Work. Blum and Speece (*47*) carried out a detailed study of toxicity in different types of systems for the purpose of enabling the use of the broad multispecies data. The results ranged from good to excellent but, in general, the findings suggest that interspecies use of toxicity data is possible. These workers also urge caution in the use of such data.

Conclusions

The model developed here from data developed on three substantially different kinds of organisms indicate that data from one species can be used to estimate the biological response expected from other species. Obviously, care must be taken in applying this concept. It must be reemphasized that the model developed here is not considered broad based enough to be used for predictive purposes. On the other hand, the model does explain over 70% of the variance which is very good for biological systems. There can be no questions that the three test systems studied here can be modeled using descriptors similar or identical to those used for the individual species study.

More work is badly needed to extend the concept and probe the limits of multispecies data applicability. Work in the area of study methods is as important as studies of specific data sets. The present work offers one approach to the study of multispecies data. Better methods for describing the molecules under study are also needed.

Many hazardous wastes can be removed from areas or sites where remediation is comparatively easy. Sediments are one example where such is not the case and remediation represents a major challenge.

Also, the impact of uncouplers on waste water treatment processes can be overhwhelming from an operational point of view and ultimately from the standpoint of effluent quality. Industry, in general, should examine their industrial processes and exclude such materials to the extent possible.

Literature Cited

(1) Okey, R. W.; Stensel, H. D. *Toxicol. Environ. Chem.* **1993**, 40, 285.
(2) Beltrame, P.; Beltrame, P. L.; Carniti, P. *Chemosphere* **1984**, 13, 3.
(3) Beltrame, P.; Beltrame, P. L.; Carniti, P.; Lanzetta, C. *Chemosphere* **1988**, 17, 235.
(4) Rich, L. G.; Yates, O. W., Jr. *J. Applied Micro.* **1955**, 3, 95.
(5) Schneider, C. G. Ph.D. Dissertation, Vanderbilt University, Nashville, TN, 1987.
(6) Maloney, P. C. *J. Membrane Biol.* **1982**, 67, 1.
(7) Chudoba, P.; Morel, A; Capdeville, B. *Environ. Tech.* **1992**, 13, 761.
(8) Harold, F. M. *The Vital Force: A Study of Bioenergetics.* W. H. Freeman and Company, New York, NY, 1986.
(9) Fruton, J. S.; Simmonds, S. *General Biochemistry.* John Wiley and Sons, Inc., New York, NY, 1958.
(10) (Servizi, J., personal communication, 1967.)
(11) Metcalf, R. L. *Organic Insecticides.* Interscience Publ., Inc., New York, NY, 1955.
(12) Clowes, G. H. A.; Keltch, A. K.; Strittmatter, G. F.; Walters, C. P. *J. Gen. Physiol.* **1950**, 33, 555.
(13) Clowes, G. H. A.; Krahl, M. E. *J. Gen. Physiol.* **1936a**, 20, 145.
(14) Clowes, G. H. A.; Krahl, M. E. *J. Gen. Physiol.* **1936b**, 20, 173.
(15) Cajina-Quezada, M.; Schultz, T. W. *Aquatic Toxicol.* **1990**, 17, 239.
(16) Ravanel, P.; Taillandier, G.; Tissut, M. *Ecotoxicol. Environ. Safety* **1989**, 18, 1.
(17) Okey, R. W.; Stensel, H. D. *Wat. Environ. Res.* **1993a**, 65, 772.
(18) Okey, R. W.; Stensel, H. D. *A Biodegradability Model and a Three-Dimensional View of Biorefractivity*, **1993b**, Paper presented at WEF Annual Conf., Anaheim, CA, accepted for publication in *Wat. Environ. Res.*, 1994.
(19) Ecosoft, Inc. *Microstat-II, Interactive Statistical Software System*, Indianapolis, IN, 1990.
(20) SPSS, Inc. *Computer Software Program*, Chicago, IL, 1990.
(21) Belsley, D. A.; Kuh, E.; Welsch, R. E. *Regression Diagnostics: Identifying Influential Data and Sources of Collinearity*. John Wiley and Sons, New York, NY, 1980.
(22) Sablijić, A. *Graph III. IBM PC Computer Program*, Institute Ruder Bošković, Zagreb, Croatia, Yugoslavia, 1989.
(23) Jandel Scientific. *SigmaPlot, Scientific Graph System*, San Rafael, CA, 1992.
(24) Chatterjee, S.; Price, B. *Regression Analysis by Example*, 2nd Ed., John Wiley and Sons, Inc., New York, NY, 1991.
(25) Cook, R. D. *Technometrics* **1977**, 19, 15.
(26) Desai, S. M.; Govind, R.; Tabak, H. H. *Environ. Toxicol. Chem.* **1990** , 9, 473.
(27) James, F. C.; McCulloch, C. E. *Annu. Rev. Ecol. Syst.* **1990**, 21, 129.

(28) Kier, L. B.; Hall, L. H. *Molecular Connectivity in Structure-Activity Analysis*. John Wiley and Sons, Inc., New York, NY, 1986.

(29) Okey, R. W. Ph.D. Dissertation, University of Washington, Seattle, WA, 1992.

(30) Hansch, C.; Leo, A. *Substituent Constants for Correlation Analysis in Chemistry and Biology*. John Wiley and Sons, New York, NY, 1979.

(31) Randic, M. *AIChE J*. 1975, 97, 6609.

(32) Pitter, P.; Chudoba, J. *Biodegradability of Organic Substances in the Aquatic Environment*. CRS Press, Ann Arbor, MI, 1990.

(33) Loch-Caruso, R.; Corcos, I. A.; Trosko, J. E. *J. Toxicol. Environ. Health*, 1991, 32, 33

(34) Jeanne, N.; Dazy, A. C.; Moreau, A. *Hydrobiologica* 1993, 252, 245.

(35) Webster, E. A.; Gadd, G. M. *Environ. Toxicol. Wat. Qual: An Intl. J*. 1992, 7, 189.

(36) Neilson, A. H.; Allard, A.-S.; Hynning, P.-A.; Remberger, M. *Toxicol. Environ. Chem*. 1991, 30, 3.

(37) Wulff, F.; Rahm. L. *Mar. Pollut. Bull*. 1993a, 26, 272.

(38) Wulff, F.; Rahm, L.; Jonsson, P.; Grydsten, L.; Ahl, T.; Granmo, A. *AMBIO*, 1993b, 22, 27.

(39) Jonsson, P.; Rappe, C.; Kjeller, L.-O.; Kierkegaard, A.; Håkanson, L.; Jonsson, B. *AMBIO* 1993, 22, 37.

(40) Garden, S.; Tseng, T. *Chemosphere* 1990, 20, 1695.

(41) Carr, A. W.; McCracken, R. O.; Stillwell, W. H. *J. Parasitol*. 1993, 79, 198.

(42) Asami, T.; Baba, M.; Yoshida, S. *Biosci. Biotech. Biochem*. 1993, 57, 350.

(43) Muzafarov, E. N.; Zolotareva, E. K. *Biochem. Physiol. Pflanzen*. 1989, 184, 363.

(44) Miyoshi, H.; Tamaki, M.; Harada, K.; Murata, H.; Suzuki, M.; Iwamura, H. *Biosci. Biotech. Biochem*. 1992, 56, 1776.

(45) Christensen, D. P.; Hutkins, R. W. *Appl. Environ. Microbiol*. 1992, 58, 3312.

(46) Okereke, A.; Montville, T. J. *Appl. Environ. Microbiol*. 1992, 58, 2463.

(47) Blum, D. J. W.; Speece, R. E. *Res. J. WPCF* 1991, 63, 198.

RECEIVED June 29, 1995

INDEXES

Author Index

Affiliation Index

Subject Index

A

Acid mine drainage, production by sulfide minerals, 196
Actinomycetes, hydrocarbon utilization, 254–256,261
Activated sludge, uncoupling, 301–303
Adjuvants, types, 224
Air–nitric acid destructive oxidation of organic wastes
advantages, 162
carbon balance, 159,160t
experimental description, 158–159
oxidation rate vs. temperature, 159–161
palladium and rhodium as catalysts, 161
Alkane metabolism by *Rhodococcus erythropolis*
biochemical analysis, 255t–257
experimental description, 253–254
growth
of mutants on alkane metabolic intermediates, 259–262
on predicted alkane metabolic intermediates, 258–259,261
hydrocarbon utilization by Actinomycetes, 254,256,261
isolation of mutants defective in alkane metabolism, 257–258
previous studies, 252
steps, 262
2-Amino-4-hydroxy-6-(isopropylamino)-*s*-triazine, role in degradation of atrazine, 175–193
Amorphous iron hydroxide, treatment of industrial wastewater using antimony adsorption, 64–73
Anionic metal adsorption onto iron oxides
effect of concentration, 67,69,70f
effect of pH, 67,68f
effect of temperature, 67,69,70f

Anionic metal adsorption onto iron oxides—*Continued*
experimental description, 66–67
experimental materials, 66
industrial examples, 72–73
rate of reaction, 67,70f
sulfate interference, 69,71–72
Antibacterials, uncoupling of oxidative phosphorylation, 307
Antimony
adsorption onto metal oxides, 64–73
applications, 64
removal from solutions using lime or sodium hydroxide, 64–65
Applicable Relevant and Appropriate Standards, vitrification technologies, 114–118
Arbacia egg, uncoupling, 301,304t
Arsenic, adsorption onto metal oxides, 65
Atrazine
photocatalytic destruction using TiO_2 mesh, 174–193
use, 175

B

Babcock & Wilcox cyclone furnace, description, 105,106f,108–109
Basic extraction sludge treatment, applications, 229–230
Bench-scale chemical treatability of Berkeley pit water
amount of acid production, 196
chemistry, composition of water, 197
copper concentration before neutralization, 207–208
large-scale two-stage neutralization, 205–206

Production: Susan Antigone
Indexing: Deborah H. Steiner
Acquisition: Rhonda Bitterli & Michelle D. Althuis
Cover design: Neal Clodfelter

Printed and bound by Maple Press, York, PA

Bestsellers from ACS Books